Synthesis and Applications
of DNA and RNA

Synthesis and Applications of DNA and RNA

Edited by

Saran A. Narang

Division of Biological Sciences
National Research Council of Canada
Ottawa, Ontario, Canada

1987

ACADEMIC PRESS, INC.

Harcourt Brace Jovanovich, Publishers

Orlando   San Diego   New York   Austin
Boston   London   Sydney   Tokyo   Toronto

ACADEMIC PRESS, INC.
Orlando, Florida 32887

United Kingdom Edition published by
ACADEMIC PRESS INC. (LONDON) LTD.
24–28 Oval Road, London NW1 7DX

Library of Congress Cataloging in Publication Data

Synthesis and applications of DNA and RNA.

    Includes index.
    1. Nucleic acids—Synthesis. 2. Nucleic acids—
Analysis.   3. Molecular genetics.   I. Narang, Saran A.
QP620.S96   1987   547.7'9   86-25936
ISBN 0—12—514030—4 (alk. paper)

PRINTED IN THE UNITED STATES OF AMERICA

87 88 89 90   9 8 7 6 5 4 3 2 1

To my inspiration Sandhya, my wife

Contents

Preface

About a billion years ago, Nature drafted the blueprints of life describing our roots, heritage, and the process of evolution. The language was DNA, its alphabet A, C, G, and T. For centuries mankind has been curious about learning and decoding the language into meaningful information. In spite of tedious and painstaking efforts to understand the chemistry and biology of DNA in the past century, unexpected complexities continue to develop.

The ability to recreate the man-made version of the DNA molecule offers a unique opportunity to learn about this molecule of life. During the last decade the art of chemical synthesis has reached a maturity which will play an invaluable role in advancing our knowledge. Many who have played significant roles in the development of this field have contributed to this book, whose main goals are to stimulate the interest of a younger generation of scientists and to share our excitement in struggling through the voyage of developing the synthetic route to DNA.

It is a pleasure to thank my colleague Dr. Roland Brousseau for his help and advice during the preparation of this book and Dr. Wayne A. Jones for his careful reading and editorial correction of the manuscript. The main theme of this book, developed in my laboratory, would not have been possible without the driving force and encouragement of my wife Sandhya and my daughter Monica Ajoo, who have filled my life with joy.

Saran A. Narang

Introduction

*There can be excitement, adventure, challenge,
and there can be great art in organic synthesis.*
Robert Woodward

To organic chemists, this statement is reason enough to justify their efforts to advance scientific knowledge in their particular fields. In 1828, Wohler's spirit of adventure led him to the crucial experiment in which ammonium cyanate was converted to urea. This simple experiment, which unified biology and chemistry, forever changed the fate of the world. Released from dogmatic fetters about *force vitale,* chemistry quickly became one of the driving forces of the industrial revolution, adding immeasurably to its impetus.

During the past hundred years, chemists have synthesized several million different compounds, some related to natural products and some of which had never before existed. The introduction of spectroscopic techniques such as infrared in the 1950s and nuclear magnetic resonance in the 1960s led to a golden age of organic synthesis, crowned by achievements such as Robert Woodward's synthesis of vitamin B_{12}. A branch of science which can accomplish the synthesis of chemical compounds possessing extremely complex structures is clearly in a very healthy state. These outstanding syntheses belong to the great cultural heritage of our age, no less so than anything produced in literature or the arts.

Inevitably scientists continued to probe the nature of living cells at the molecular level. The major achievement of this approach is undoubtedly Watson and Crick's discovery of the double-helix structure of DNA in 1953, a discovery which began a golden era in molecular biology. Knowledge of how genetic information flows from DNA to mRNA to protein and the elucidation of the genetic code and the confirmation of its quasi-universality have also come about, at least in part, through a more rigorous application of the molecular approach to living systems.

These discoveries have wrought a radical change in our ideas about microbes. We now view them as tiny biochemical factories and their long familiar actions as highly sophisticated chemical transformations. A bacterium such as *Escherichia coli* reproduces in 20 minutes under optimal growth conditions—a new living cell is synthesized within one-third of an hour. Not only must the bacterium replicate accurately (one base pair mismatch can be lethal) the 2 million bases of its genome, it must also synthesize the correct amounts of over 3500 different proteins, and all within 20 minutes. The elucidation of the extremely precise chemical control mechanisms involved will remain a fruitful field of study for scientists for decades to come.

After 1953, it was clear that a rich field of DNA synthesis awaited ambitious chemists, and H. G. Khorana was one to undertake the challenge. Knowing DNA to be a macromolecule, he reasoned that with chemical reagents alone meaningful lengths of it could be synthesized. By using equally well-known enzymes and chemical reagents, he further closed the gap between biology and chemistry, made a determinant contribution toward the elucidation of the genetic code, and showed the way in which gene synthesis could eventually be undertaken.

In the 1970s, tremendous progress in the development of rapid chemical synthesis of polynucleotides, DNA sequence analysis, and recombinant DNA technology led to a genetic engineering revolution, one that has the potential to touch nearly every phase of human life. Already recombinant DNA production of human insulin, growth hormone, and interferon has become a reality, and these are but small achievements in the face of things to come.

As the twenty-first century approaches, we are faced with another challenge of code elucidation, this one pertaining to protein folding. Globular proteins consist of linear polypeptide chains which are synthesized from 20 different amino acids. During or after synthesis these chains fold spontaneously to an exact three-dimensional structure. The spontaneity of the folding is regarded as a generally valid principle because it can be demonstrated in renaturing experiments with several proteins. Yet the mathematics describing this process is staggering. For example, folding of a chain of 100 amino acid residues requires the adjustment of about 300 parameters. Even if each parameter is restricted to only two values, some 2^{300} configurations become possible. Random adoption of each configuration until the chain stumbles onto the right one would require a time exceeding the age of the earth even if only 10^{-13} seconds were assumed for each trial. Hence, the parameter values must be run through in a directed manner, i.e., there must be a defined folding pathway so that little time is lost with trial-and-error folding.

There is hope that this mechanism can be elucidated and reproduced by using an integrated approach involving modern computational tools, extensive gene synthesis, and vast amounts of data drawn from X-ray crystallography, protein chemistry, protein spectroscopy, and genetic engineering technology. The knowledge gained will eventually allow scientists to design custom-made enzymes, hormones, and structural proteins, and will open the new era of protein engineering.

Saran A. Narang

1

Development of Chemical Synthesis of Polynucleotides

SARAN A. NARANG

Division of Biological Sciences
National Research Council of Canada
Ottawa, Ontario, Canada K1A OR6

I. INTRODUCTION

Our roots and the history of evolution are written in terms of codes in the base sequence of DNA. These facets of DNA have imposed the coding concept on the field of modern biology. Thus, the goal of the study of DNA is to decode and understand the biological information contained in DNA sequences. Progress in the last 20 years has made us aware of the role of various DNA sequences. We can now read and translate DNA sequence into protein sequence and regulatory signals that control the gene activity. The use of synthetic, tailor-made DNA has played a key role in understanding the genetic code (Khorana *et al.*, 1966) and various regulatory signals such as the operator, promoter, ribosomal binding sites, enhancers, transposable elements, and homeo-box. The synthetic approach not only provides a final proof of the roles of various DNA sequences but also offers an opportunity for further improvement in function for practical application. The application of synthetic genes, linkers, primers, and probes has become a powerful tool in the cloning, sequencing, and isolation of genomic DNA. In the following sections, I have outlined how various stages of development in the chemical synthesis of polynucleotides helped to achieve the present status of this field.

1

Synthesis and Applications
of DNA and RNA

II. HISTORICAL BACKGROUND

The present foundation of synthetic work was laid by the prolonged and intensive efforts of Todd and Khorana during the 1950s. Although Michelson and Todd (1955) were the first to correctly link dinucleotides by condensing 5'-acetylthymidine 3'-benzylphosphorochloridate with 3'-acetylthymidine, nonetheless their method has limitations. The overall yield of the final product was low because of various side reactions. Thus, this approach was not useful in the synthesis of higher oligo- or polynucleotides. The search continued in Todd's laboratory for a superior (that is, milder and more effective) chemical phosphorylating method. In 1953, Khorana and Todd described the use of a different phosphorylating reagent, dicyclohexylcarbodiimide (DCC). Todd and associates published a further study later in that same year using DCC for the synthesis of coenzymes; (Christie *et al.*, 1953) but it was Khorana who used this reagent for polynucleotide synthesis. Between 1956 and 1958, Khorana and co-workers in Vancouver employed this reagent extensively to establish a phosphodiester approach of oligonucleotide synthesis which became routine for the next two decades (Gilham and Khorana, 1958). In spite of the fact that this approach was laborious and time-consuming, they did succeed in accomplishing the total synthesis of a biologically active tyrosine tRNA gene (Khorana, 1979).

By the mid-1960s, other organic chemists, such as Letsinger, Reese, Eckstein, and Cramer, reinvestigated the phosphotriester approach first introduced by Michelson and Todd (1955), in which the three phosphate bonds were masked during synthesis. Although the synthesis of oligomers up to octamers was obtained in a reasonable yield, it did not attract much attention as compared to the established approach yielding a phosphodiester. Again it was the development of new phosphorylation and coupling reagents such as triazole and tetrazole derivatives by Narang (Katagiri *et al.*, 1974) which established the phosphotriester method on a firm footing as an alternative to the phosphodiester approach. Using this method, the first biologically active genetic element, the *lac* operator DNA, was synthesized and cloned (Bahl *et al.*, 1976). Subsequently, genes encoding other chemically important proteins such as insulin (Goeddel *et al.*, 1979) and interferons (Edge *et al.*, 1981) were synthesized via the same technique.

In 1976, Letsinger and Lunsford introduced a phosphite triester approach using phosphodichloridites to link nucleosides in a much shorter time than before; this was further modified to the phosphoramidite approach by Matteucci and Caruthers in 1981. This approach has been the basis for several commercial solid-phase DNA synthesizers.

III. CHEMICAL SYNTHESIS

The fundamental objective of oligonucleotide synthesis is the formation of $3' \rightarrow 5'$ phosphate linkages between nucleoside units. This has been achieved by the three reactions outlined in Fig. 1.

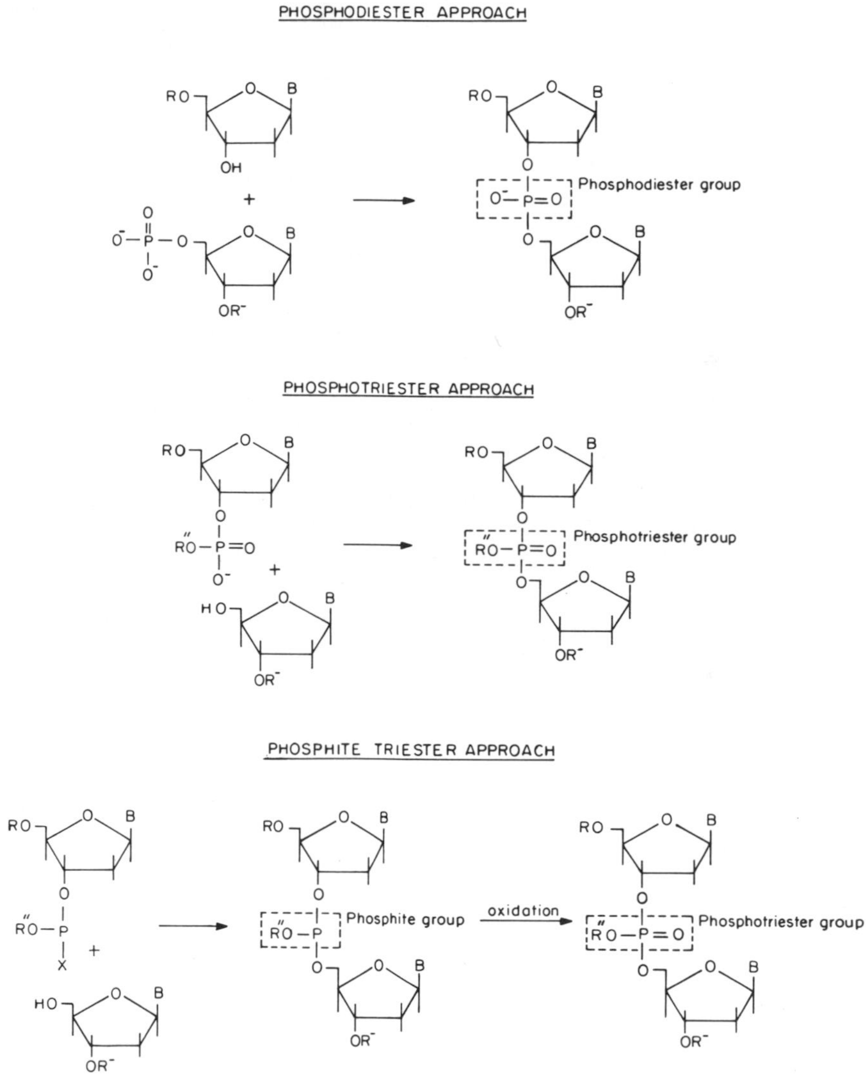

PHOSPHODIESTER APPROACH

PHOSPHOTRIESTER APPROACH

PHOSPHITE TRIESTER APPROACH

Fig. 1. Chemical approaches for the synthesis of a polynucleotide.

A. Phosphodiester Approach

The phosphodiester approach involves the condensation of a 5'-phosphomonoester group of a suitably protected nucleoside with the 3'-hydroxyl function of another nucleotide in the presence of a coupling reagent such as dicyclohexylcarbodiimide (DCC) or aryl sulfonyl chloride. Because the product contains phosphodiester linkages, this approach is termed the phosphodiester method. Because the starting material and product are ionic in nature, ion-exchange column chromatography has to be used extensively in the purification of products. The yield of the larger size fragment decreases because of the instability of the internucleotidic phosphodiester bonds. This can only be overcome by using a large excess of the incoming unit which makes this approach uneconomical and laborious.

B. Phosphotriester Approach

Some of the problems inherent in the phosphodiester method, such as low yield and laborious ion-exchange chromatographic purification techniques, can be solved if the third dissociation of phosphate is masked to create a neutral molecule. The more standard organic chemistry techniques can be employed to isolate the product. This is known as the phosphotriester approach because of the presence of the triester–internucleotide bond in the intermediate product. To mask the diester, various synthetic routes have been investigated. Letsinger and Mahadevan (1965) reported the two-step synthesis of a dinucleotide. This involves phosphorylation of 5'-protected thymidine with β-cyanoethyl phosphate and mesitylenesulfonyl chloride followed by condensation with a second nucleoside in the presence of a coupling reagent, triisopropylbenzenesulfonyl chloride. The dinucleotide was isolated on silical-gel chromatography in 64% yield. Reese and Saffhill (1968) used a similar approach employing phenylphosphodichloridate as the phosphorylating reagent; this was replaced by o-phenyl phosphate in 1971 (van Boom et al., 1971) because of its base-labile property. In 1973, Reese et al. succeeded in achieving a large-scale synthesis of an octamer of the thymidine series.

Eckstein and Rizk (1967) employed a modified strategy of phosphotriester synthesis. It involved the condensation of 5'-protected thymidine 3'-(2,2,2-trichloroethyl)phosphate with 3'-acetylthymidine in the presence of triisopropylbenzenesulfonyl chloride. In 1973, Catlin and Cramer employed a similar strategy and succeeded in achieving the synthesis of a tetranucleotide. In the same year Narang and co-workers (Itakura et al, 1973) also succeeded in achieving the synthesis of the hexadecanucleotide of the thymidine series by block condensation.

C. New Phosphorylating and Condensing Reagents

The development of DCC and aryl sulfonyl chloride as condensing reagents by Khorana and co-workers (Smith *et al.*, 1958); Jacob and Khorana, 1964) has played a significant role in the development of the phosphodiester approach. These reagents were found to be unsatisfactory in phosphotriester methodology because aryl sulfonyl chloride caused extensive sulfonation and DCC could not activate the phosphodiester functional group. In 1974, Narang and associates overcame this problem by introducing triazole derivatives of phosphoryl and aryl sulfonyl moieties. This was followed by tetrazole (Stawinski *et al.*, 1976) and 3-nitro-1,2,4-triazole (Reese and Ubasawa, 1980). Although these reagents had been studied extensively, in 1982 Efimov *et al.* reported that aryl sulfonyl chloride in the presence of *N*-methylimidazole can catalyze the formation of the phosphotriester bond in neutral organic solvents. Recently, they introduced N-substituted derivatives of pyridine 4-oxide and quinoline *N*-oxide in conjunction with aryl sulfonyl chloride which led to a dramatic increase in the rate of phosphotriester formation (less than 1 min) (Efimov *et al.*, 1985). This is comparable to the rate of phosphate bond formation (*vide infra*) in phosphoramidite chemistry.

D. Modified Phosphotriester Approach

The basic principle of this approach is to start the synthesis from 5'-*O*-dimethoxytritylnucleoside 3'-triester phosphate as outlined in Fig. 2. On acid treatment the 5'-hydroxyl group is made free, and in base the β-cyanoethyl group is removed from 3'-triester phosphate. The coupling of these two components generates a fully protected dinucleotide containing the 3'-phosphotriester group. Since each intermediate oligonucleotide contains a fully masked 3'-phosphate group, the necessity for phosphorylation before each condensation step was eliminated, thus simplifying this approach. Using this approach, block condensation led to the synthesis of large-size molecules that on complete deblocking yielded oligonucleotides containing normal phosphodiester bonds.

E. Phosphite Triester Approach

The basic feature is the linking of nucleosides through a phosphite internucleotide bond, which on subsequent oxidization generates a phosphotriester bond. Letsinger and Lunsford (1976) developed this approach using *O*-chlorophenylphosphodichloride for the phosphorylation of the 3'-hydroxyl function of the nucleoside, followed by coupling to the second nucleoside. One of the drawbacks of this method was the low temperature

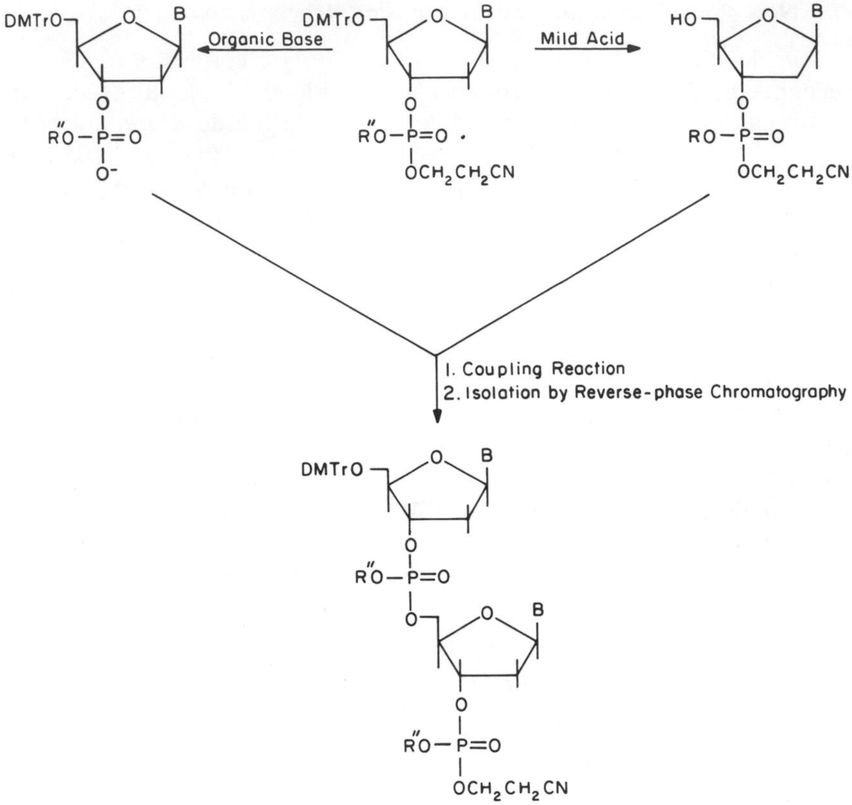

Fig. 2.   Modified phosphotriester approach.

reaction condition ($-78°C$). McBride and Caruthers (1983) overcame this problem by preparing morpholine and diisopropylphosphoramidite derivatives of the nucleoside which were nonhydroscopic and stable at room temperature as outlined in Fig. 3.

IV.   CONCLUDING REMARKS

During the last decade, great progress in the methodology of the chemical synthesis of DNA, including automatic DNA synthesizers, has been made in terms of speed, yield, and length of fragments. These fragments, designed to form a DNA duplex, can be readily joined to a plasmid and cloned in host cells by a one-pot ligation–cloning procedure. The gene synthesis process—starting from chemical fragments to gene assembly,

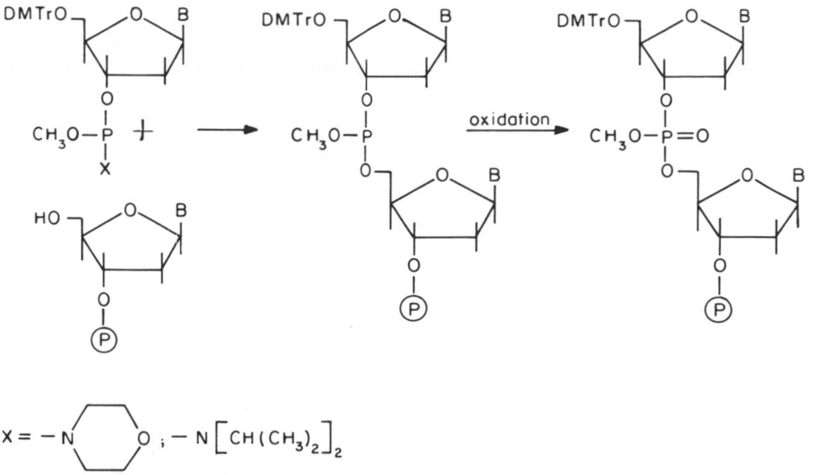

$$X = -N\underset{\smile}{\overset{\frown}{\bigcirc}}O \; ; \; -N\left[CH(CH_3)_2\right]_2$$

Fig. 3. Phosphoramidite approach.

cloning, selection, and sequence confirmation—can be accomplished within a couple of weeks.

Equipped with this powerful tool, organic chemistry has entered a new era. Soon it will be possible to learn the structure–function relationship of a protein and its folding mechanism through a DNA-directed protein synthesis approach. It is predicted that by the end of the 20th century it will be possible to design and synthesize a new enzyme capable of an expected function. Thus we will reach a stage in which the synthesis of a macromolecule such as an enzyme will be easier than a small-sized molecule because the synthesis of a small molecule requires more than one enzyme for a catalyst. Is it not paradoxical?

REFERENCES

Bahl, C. P., Wu, R., Itakura, K., and Narang, S. A. (1976). *Proc. Natl. Acad. Sci. U.S.A.* **73**, 91–94.

Catlin, J. C., and Cramer, F. (1973). *J. Org. Chem.* **38**, 245–250.

Christie, S. M. H., Elmore, D. T., Kenner, G. W., Todd, A. R., and Weymouth, F. J. (1953). *J. Chem. Soc.*, 2947–2953.

Cusak, N. J., Reese, C. B., and van Boom, J. H. (1973). *Tetrahedron Lett.*, 2209–2212.

Eckstein, F., and Rizk, I. (1967). *Angew. Chem., Int. Ed. Engl.* **6**, 695–696.

Edge, M. D., Greene, A. R., Heathcliffe, G. R., Meacock, P. A., Schuch, W., Scanlon, D. B. Atkinson, T. C., Newton, C. R., and Markham, A. F. (1981). *Nature (London)* **292**, 756–761.

Efimov, V. A., Reverdatto, S. V., and Chakhmakhcheva, O. G. (1982). *Nucleic Acids Res.* **10**, 6675–6694.

Efimov, V. A., Chakhmakhcheva, O. G., and Ovchinnikov, Yu. A. (1985). *Nucleic Acids Res.* **13**, 3651–3666.

Gilham, P. T., and Khorana, H. G. (1958). *J. Am. Chem. Soc.* **80**, 6212–6222.

Goeddel, D. V., Kleid, D. G., Bolivar, F., Heynekar, H. L., Yansura, D. C., Crea, R., Hirose, T., Kraszewski, A., Itakura, K., and Riggs, A. D. (1979). *Proc. Natl. Acad. Sci. U.S.A.* **76**, 106–110.

Itakura, K., Bahl, C. P., Katagisi, N., Michniewiez, J. J., Wightman, R. H., and Narang, S. A. (1973). *Can. J. Chem.* **51**, 3649–3651.

Jacob, J. M., and Khorana, H. G. (1964). *J. Am. Chem. Soc.* **86**, 1630–1635.

Katagiri, N., Itakura, K., and Narang, S. A. (1974). *Chem. Commun.,* pp. 325–326.

Khorana, H. G. (1979). *Science* **203**, 614–625.

Khorana, H. G., and Todd, A. R. (1953). *J. Chem. Soc.,* pp. 2257–2260.

Khorana, H. G., Büchi, H., Ghosh, H., Gupta, N., Jacob, T. M., Kössel, H., Morgan, R. A., Narang, S. A., Ohtsuka, E., and Wells, R. D. (1966). *Cold Spring Harbor Symp. Quant. Biol.* **31**, 39–49.

Letsinger, R. L., and Lunsford, W. B. (1976). *J. Am. Chem. Soc.* 3655–3661.

Letsinger, R. L., and Mahadevan, V. (1965). *J. Am. Chem. Soc.* **87**, 3526–3527.

McBride, L. J., and Caruthers, M. H. (1983). *Tetrahedron Lett.* **24**, 245–252.

Matteucci, M. D., and Caruthers, M. H. (1981). *J. Am. Chem. Soc.* **103**, 3185–3191.

Michelson, A. M., and Todd, A. R. (1955). *J. Chem. Soc.,* pp. 2632–2638.

Reese, C. B., and Saffhill, R. (1968). *Chem. Commun.,* pp. 767–768.

Reese, C. B., and Ubasawa, A. (1980). *Nucleic Acids Symp. Ser.* **7**, 5–21.

Smith, M., Moffatt, J. G., and Khorana, H. G. (1958). *J. Am. Chem. Soc.* **80**, 6204–6212.

Stawinski, J., Hozumi, T., and Narang, S. A. (1976). *Can. J. Chem.* **54**, 670–672.

van Boom, J. H., Burgers, P. M. J., Owen, G. R., Reese, C. B., and Saffhill, R. (1971). *Chem. Commun.* pp. 869–871.

2

DNA Synthesis on Solid Supports and Automation

BRUCE E. KAPLAN
KEIICHI ITAKURA
Department of Molecular Genetics
City of Hope Research Institute
Duarte, California 91010

I. INTRODUCTION

Inspired by the ability of living organisms to synthesize proteins and DNA from monomeric units, chemists sought to duplicate—if only in a rudimentary fashion—these products *in vitro*. Knowing that the exquisitely precise and rapid enzymatic systems that organisms use for the synthesis of proteins and DNA were beyond the capabilities of chemists, nonenzymatic methods were sought for the synthesis of oligopeptides and oligonucleotides. The concept of using a polymer as an "inert" support for an organic synthesis was first proposed in 1962 by Merrifield.[1] This seminal principle was exploited rapidly for the synthesis of oligonucleotides by Letsinger and Mahadevan[2] in 1965. Another 9 years were to elapse until the synthesis of a heptanucleotide of diverse bases was reported by Koster, Pollack, and Cramer.[3] In 1977 Gait and Sheppard[4] reported the synthesis of several oligomers of seven to nine bases, some of which contained all four bases. The desired oligomers were not only the major components in the product mixture, but also served as specific primers for the sequencing of mRNA. About 3 years later Miyoshi, Huang, and Itakura[5] reported the synthesis of three hexadecanucleotides

9

Synthesis and Applications
of DNA and RNA

that clearly showed that almost any specific sequence could be prepared in a relatively rapid manner and easily purified. Today oligomers of up to 50 bases are synthesized routinely and polydeoxynucleotides of up to 107 bases have been synthesized.[6]

Why was the progress in polymer-supported (solid-phase) oligonucleotide synthesis so slow after the original report by Letsinger? To be able to answer this question, we have to understand the fundamental differences between solid-phase and solution-phase synthesis. In a solid-phase synthesis the purification of the desired product can only occur after the product has been released from the support. For this purification to be possible, one has to be able to separate the desired oligomer (which we shall designate of length n) from oligomers one base shorter (designated as $n - 1$) and oligomers one base longer (designated as $n + 1$). (The formation of oligomers longer than expected may occur by the coupling of a 5′-OH- nucleotide, an impurity, to the growing oligomer, followed by the coupling of a dimethoxytrityl (DMT)-nucleotide (Fig. 1). Therefore chromatographic techniques that could accomplish this result had to be developed. To simplify the chromatography it is highly desirable that the desired product be the single largest component in the product mixture. In solution-phase synthesis one frequently purifies the product after each step in the synthesis. If the yield is greater than 50%, the product molecule will be the largest component among the oligomers, and as long as we can separate the product from the starting materials synthesis can pro-

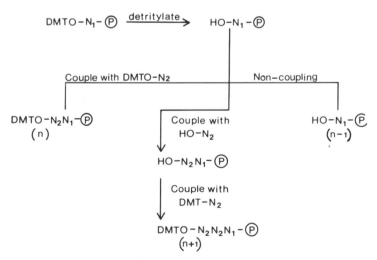

Fig. 1. Solid-phase synthesis of DNA. ⓟ, Polymer support; DMT, dimethoxytrityl; N, coupling unit.

TABLE I
Coupling Yield per Step

Yield per step (%)	Length of oligomer	Overall yield (%)
70	6-mer	11.7
80	10-mer	10.7
90	20-mer	10.9

ceed to the next step. Since each step in a solution-phase synthesis begins with an oligomer purified from the preceding step the condensation yields need only be moderate for the final synthesis to be successful. Solid-phase synthesis of oligonucleotides requires much higher yields at each step in the synthesis for the product to be isolatable. If we assume that a 10% overall yield is required for the isolation of the product oligomer, then the relationship between yield per step and length is clear from Table I. This assumes that the couplings are carried out with nucleotides of unit length. The earliest successful synthesis of long oligomers was accomplished by using dimers or trimers as the coupling units and therefore reducing the number of couplings required for oligomer assembly.[5]

The assembly of an oligonucleotide requires the formation of an internucleotidic phosphate bridge between the 5'-OH of one nucleoside and the 3'-OH of another nucleoside. Three methods of forming this linkage have become known as the phosphodiester, phosphotriester, and phosphite–triester methods (Figs. 2, 3, and 4). The phosphite–triester method is discussed in the following chapter.

Although the first synthesis of an oligonucleotide on a solid support employed the phosphotriester method,[2] most of the following synthetic efforts utilized the more studied phosphodiester method. The phospho-

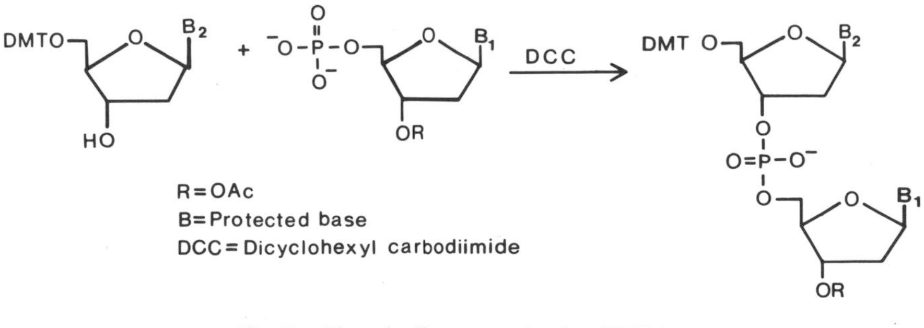

R = OAc
B = Protected base
DCC = Dicyclohexyl carbodiimide

Fig. 2. Phosphodiester synthesis of DNA.

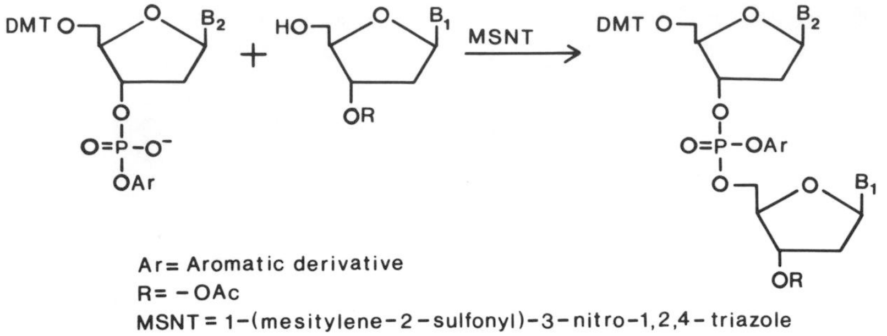
DMTO
B2
HO
B1
MSNT
DMT O
B2
O=P-O⁻
OAr
OR
O=P-OAr
B1
OR
Ar= Aromatic derivative
R= -OAc
MSNT = 1-(mesitylene-2-sulfonyl)-3-nitro-1,2,4-triazole

Fig. 3. Phosphotriester synthesis of DNA.

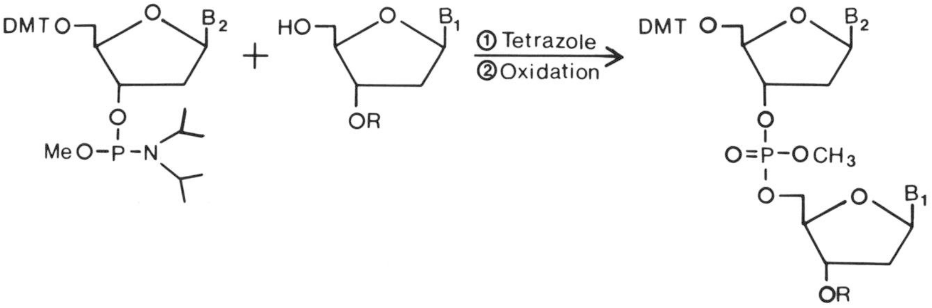
DMTO
B2
HO
B1
① Tetrazole
② Oxidation
DMT O
B2
MeO-P-N
OR
O=P-OCH3
B1
OR

Fig. 4. Phosphite–triester synthesis of DNA.

diester method, as carried out in solution, gave relatively low yields of the desired products and relatively complicated reaction mixtures. In addition, the phosphodiester method required polar solvents to solublize the charged phosphate intermediates, and, since the polymer support was most commonly polystyrene (a nonpolar polymer), a distinct incompatability existed among incoming dicharged nucleotides, the charged polynucleotides, and the support. Thus, the tertiary structure of the growing polynucleotide may have limited the access of the incoming nucleotide to the 3′-OH terminus of the polynucleotide. By the time that methods had been developed to disrupt the tertiary structure of the growing polynucleotide, using polar solvents and the development of polar solid supports,[4] the phosphotriester method had been sufficiently improved by Narang[7] and Reese[8] to supplant completely solid-phase phosphodiester synthesis and a resurgent solid-phase phosphotriester method came to the fore.[5]

II. MECHANISM OF COUPLING REACTIONS

In the polymer-supported phosphotriester method for oligonucleotide synthesis, there are many factors which can be varied from one synthesis to the next. Some of these include (1) polymer supports, (2) solvents, (3) detritylation reagents, (4) coupling reagents, (5) protecting groups on nucleotides, (6) time, and (7) temperature. Before discussing these individual variables it would be useful to understand the proposed mechanisms of the coupling reactions. Three different papers[9-11] have been published on the mechanisms of oligonucleotide synthesis by the phosphotriester method. Each study examines the problems in a somewhat different manner and in different solvents. The work of Chandrasegaran, Murakami, and Kan comes closest to one of the standard conditions for the phosphotriester method, using 1-(mesitylene-2-sulfonyl)-3-nitro-1,2,4-triazole (MSNT) in pyridine as the coupling reagent. Zaryatova and Knorre also discuss the mechanisms using mesitylene-2-sulfonyl chloride (MsCl) plus N-methylimidazole in various solvents. A holistic summary of the proposed mechanism is shown in Fig. 5.

Summary of Mechanism

The mechanism of the phosphotriester condensation can be separated into two basic steps:

First Step. Reaction of a 5′-protected nucleotide diester **(1)** with arenesulfonyl chlorides (such as MSCl), arenesulfonyl tetrazoles (such as TPSTet), or arenesulfonyl-3-nitro-1,2,4-triazoles (such as MSNT) in pyridine yields a mixture of the nucleotide–sulfonic acid mixed anhydride **(2)** [Eq. (a) in Fig. 5] and the 3′–3′ symmetrical nucleotide pyrophosphate **(3)** [Eq. (b)]. This step is shown to be catalyzed by pyridine, dimethylaminopyridine, and N-methylimidazole.[9]

Second Step. Reaction of **2** or **3** with a 5′-OH nucleoside yields the phosphotriester **(4)** [Eqs. (c,d)]. In the absence of a catalyst this reaction is very slow[11]. N-Methylimidazole acting as a nucleophilic catalyst reacts with **2** and **3** to form an intermediate which is very reactive toward 5′-OH nucleosides[9-11] [Eq. (e)]. Azoles such as imidazole, 1,2,4-triazole, 3-nitrotriazole, and tetrazole are also powerful catalysts in this second reaction step.[11]

III. POLYMER SUPPORTS

A. Preparation of Polymer Supports

Chemists have functionalized and utilized numerous types of polymer supports for oligonucleotide synthesis. Most of these supports have proved less than satisfactory, and currently only four types are commercially available, with the first nucleoside already attached to the support. These are: controlled pore glass (CPG),[12] polystyrene,[13] silica gel,[14] and

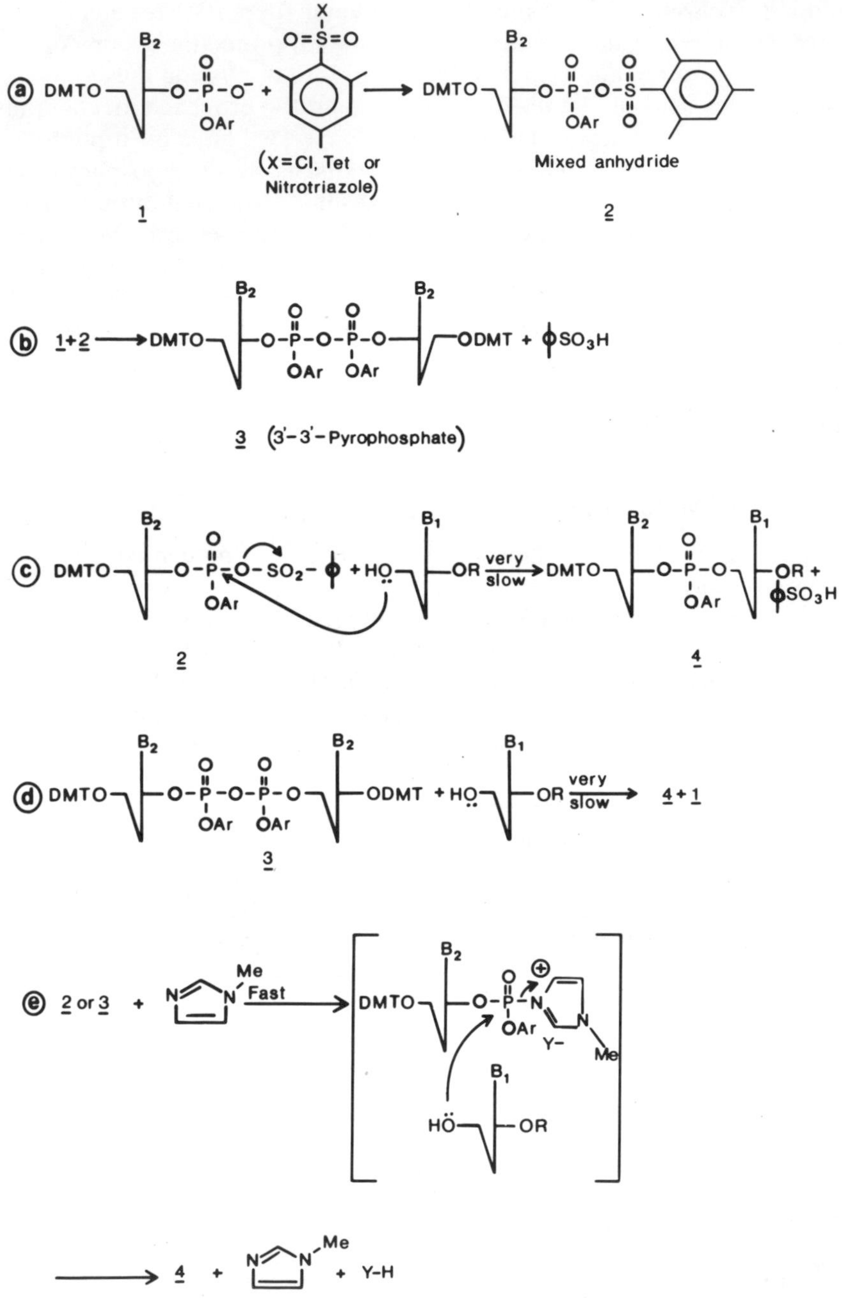

Fig. 5. Mechanism of phosphotriester coupling.

kieselguhr/polyamide.[15] Cellulose,[16] a fifth support, is frequently used but not commercially available.

1. Controlled Pore Glass

Koster[17] states that "CPG is the carrier of choice for polymer support oligonucleotide synthesis . . ." and we believe that this is true for most of the small-scale oligonucleotide syntheses that are performed on automatic synthesizers. CPGs come in a variety of pore sizes: 240, 500, 1400, and 3000 Å. The variety commonly functionalized with the first nucleoside attached is the 500 Å CPG. There are numerous functionalized CPGs; two that have been successfully used are shown in Fig. 6. The loading of the aminopropyl groups on the CPG is a function of the pore size of the CPG and varies from 20 μmoles/g for the 3000 Å CPG to 140 μmoles/g for the 240 Å CPG.[17] The accessibility of these aminopropyl groups to the nucleoside succinates are a function of the pore size. Thus, the 3000 Å CPG could only be loaded to a level of 2 μmole/g (with the nucleoside succinate) while the 240 Å CPG could be loaded to 100 μmoles/g.[17] In addition to their chemical superiority, CPGs also possess mechanical

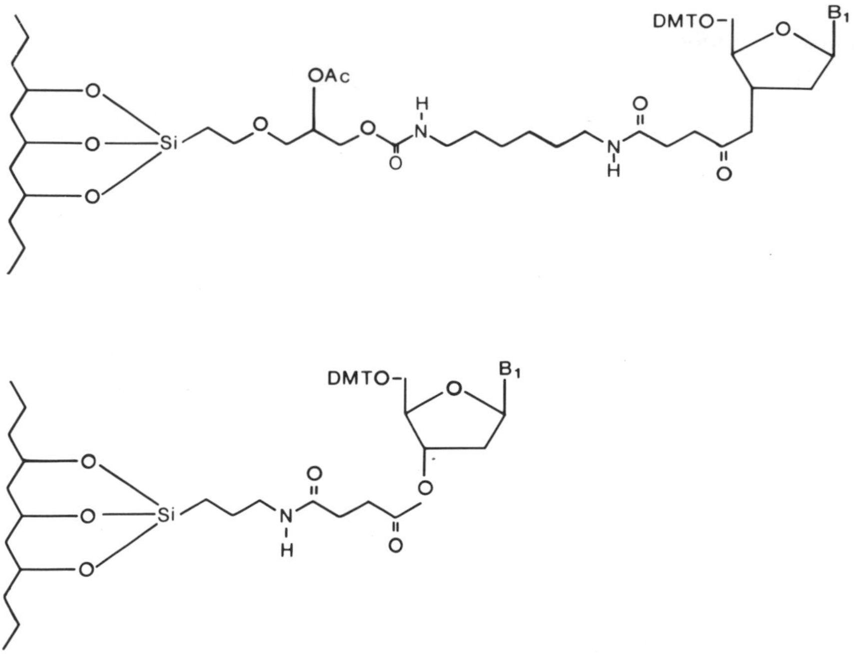

Fig. 6. Functionalized CPG supports for DNA synthesis.

properties that offer distinct advantages as a polymer support for synthesis. CPGs are not friable and thus the particles can be kept whole during the process of functionalization and synthesis. This means that the flow rates through the material are relatively constant, which makes for an easier synthesis. Most CPGs have a nucleoside loading of 20–30 μmoles/g and a density of about 0.35 g/cm^3. Therefore a 1 μmole scale synthesis requires about 50 mg of CPG and occupies about 150 μl. Since it is imperative that the CPG be completely immersed in the coupling mixture, this requires about 150–200 μl of solution. If one wishes to carry out a large-scale synthesis, the cost of these reagents gets very high. A possible solution is to employ a higher loaded solid support, such as silica gel.

2. Silica Gel

Silica gels can be functionalized by the same methods as those employed for CPGs. The loadings of the nucleoside are ~200 μmole/g[18] For larger-scale syntheses of oligonuleotides this can offer significant savings in costly reagents. The silica gel is quite friable and must be handled carefully. Furthermore small particles produced during its functionalization or during the oligonucleotide synthesis will clog the filters in the reaction vessel causing unpredictable flow rate.

3. Polystyrene

Polystyrene, the first polymer support used for oligonucleotide synthesis, is still used when a large amount of oligomer is to be synthesized. The reactions necessary to functionalize polystyrene are shown in Fig. 7. The type of polystyrene best suited to oligonucleotide synthesis is a 1% divinylbenzene cross-linked resin that has been aminomethylated to a level of ~200 μmoles/g. The nucleoside ~120 μmoles/g) is then connected to the aminomethyl polystyrene through a succinate linkage.[19]

4. Polyamide/Kieselguhr

A rather unusual polydimethylacrylamide has been prepared inside a shell of kieselguhr.[15] This material, which has been functionalized and loaded to a nucleoside level of ~110 μmoles/g, may be quite useful for large-scale syntheses of oligonucleotides. Up to the present it has not been used much outside the laboratories in which it was developed.

5. Cellulose

Although cellulose was first introduced as a support for oligonucleotide synthesis by Crea and Horn[20], it was rarely used until the filter disk method of Blocker[16] was developed. In this method small cellulose disks are functionalized and loaded with the different nucleosides. Each disk is

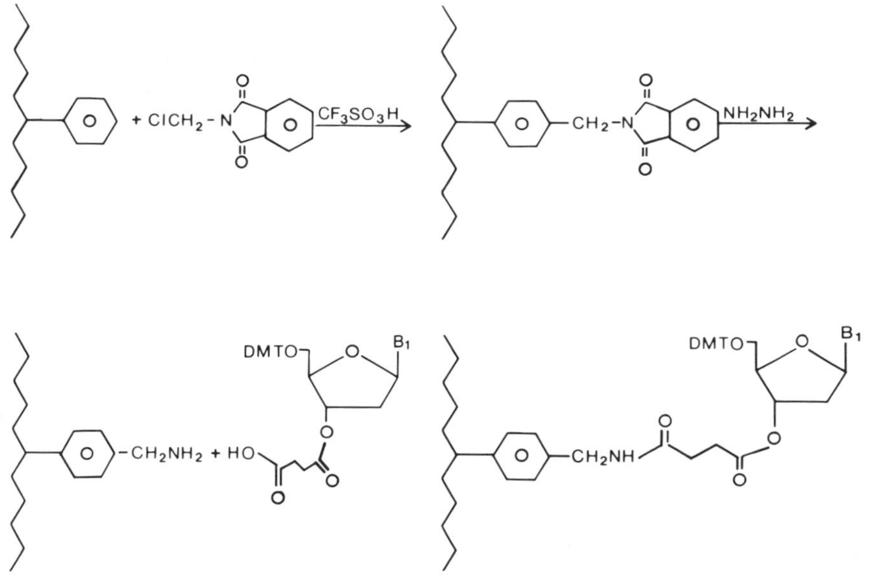

Fig. 7. Chemical reactions for the functionalization of polystyrene.

then designated as the 3' end of an oligonucleotide to be synthesized. All of the disks that are to have a particular nucleotide connected are placed together in a reaction vessel. The disks are then detritylated and coupled with the required nucleotide. After appropriate washings the disks are sorted for the next coupling reaction. This method can lead to dramatic savings when there are many oligonucleotides to be synthesized on a small scale (as when a gene is to be synthesized) (Fig. 8). Thus, the

Coupling #	Supports used	Nucleotide coupled	Products
1	A-(P), G-(P)	T	TA-(P), TG-(P)
2	T-(P), C-(P)	A	AT-(P), AC-(P)
3	TA-(P), TG-(P) AT-(P), AC-(P)	C	CTA-(P), CTG-(P) CAT-(P), CAC-(P)
4	CTA-(P), CTG-(P) CAT-(P), CAC-(P)	G	GCTA-(P), GCTG-(P) GCAT-(P), GCAC-(P)

Fig. 8. Filter disk-supported synthesis. Oligonucleotides to be synthesized: 5'-GCTA-3'; 5'-GCTG-3'; 5'-GCAT-3'; 5'-GCAC-3'. Starting materials required: 3'A-(P), 3'T-(P), 3'T-(P), 3'C-(P), 3'G-(P) where (P) symbolizes the cellulose support.

formation of four tetranucleotides requires four couplings instead of twelve. The savings get much more dramatic as the number of syntheses is increased.[21]

B. Linkage between the First Nucleoside/Nucleotide and the Polymer Support

Almost all polymer-supported oligonucleotide synthesis is directed from the 3' → 5'. One reason for choosing this direction is that the condensation between the 3'-phosphate and 5'-OH is about four times faster than that between the 5'-phosphate and the 3'-OH.[11] In almost all syntheses the 3'-OH is connected through a succinate ester linkage, which in turn is connected to the polymer support through a succinate amide linkage (Figs. 5 and 6). This mode of connecting the nucleoside to the polymer support is adequate if at the end of the synthesis the free oligonucleotide is the desired product. If an oligonucleotide with a 3'-phosphate is the desired product or if a fully protected (except for the 3'-phosphodiester terminus) oligonucleotide is required for a block synthesis, then a more versatile linkage is required. Just such linkages have been utilized by Felder[22], based on the fact that each degree of oxidation of the parent link molecule 2-(4-carboxyphenylmercapto)ethanol (CAMET) corresponds to different stabilities and cleaving conditions. The CAMET link is stable during storage, synthesis, and deprotection of phosphate esters. Upon oxidation of the sulfide to sulfoxide or sulfone, the linkage becomes labile to base via a β-elimination reaction. The 2-benzylsulfonylethyl (BSE) linkage, proposed by Balgobin[23] for oligonucleotide synthesis, has been adopted by Efimov[24] for syntheses of oligonucleotide blocks, where a complete deprotection is not required. The BSE linkage is easily cleavable with triethylamine/pyridine. (See Figs. 9 and 10.)

The idea of using a single support for the synthesis of any oligomer was first suggested by Koster[25] using the phosphodiester approach. Recently Gough[26] has demonstrated that a uridine linked to a CPG can serve as a universal support (Fig. 11). An additional benefit of this support is that it avoids connecting dA nucleosides to the polymer support. It is known that N-bz-dA is very susceptible to depurination, especially when connected to a polymer support as the nucleoside.[27]

IV. PROTECTING GROUPS USED DURING OLIGONUCLEOTIDE SYNTHESIS

The synthesis of any large molecule, such as an oligonucleotide, requires that many potential reactive centers be blocked during the synthe-

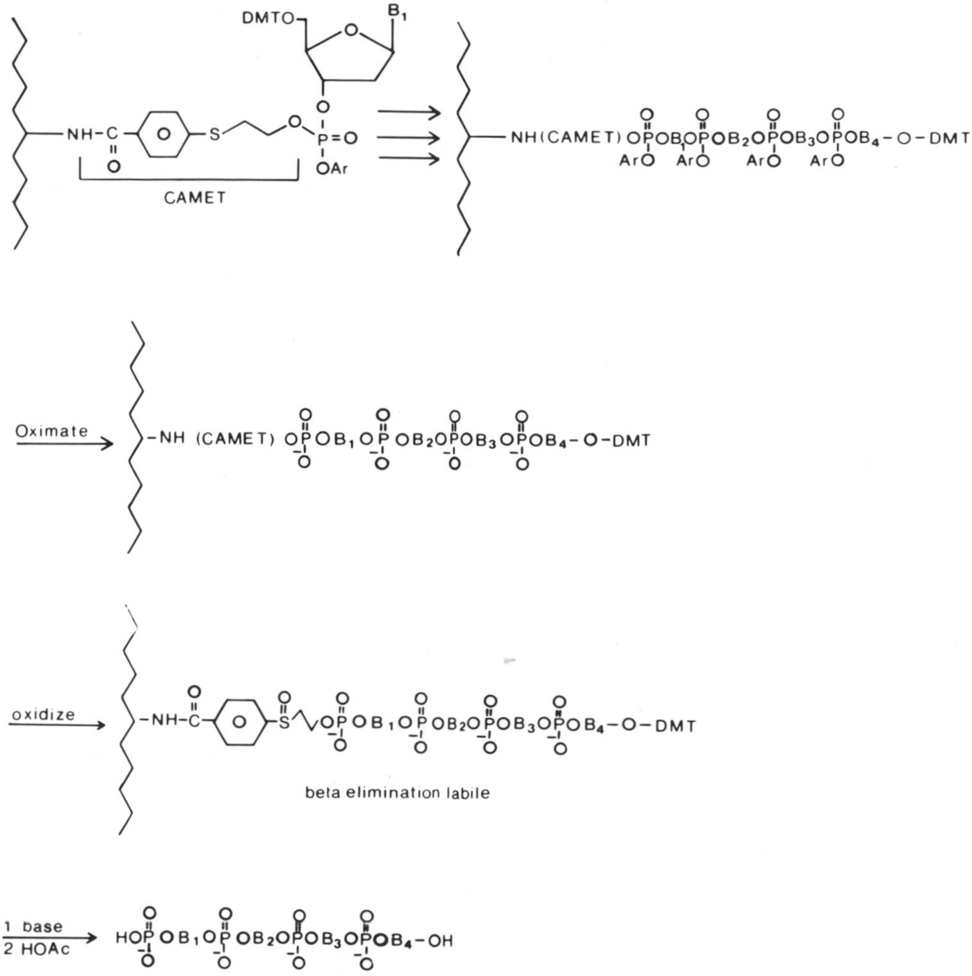

DMTO— B₁

—NH—C—⟨O⟩—S ... O—P=O / OAr

CAMET

—NH(CAMET) OPOB₀POB₂OPOB₃OPOB₄—O—DMT
ArO ArO ArO ArO

Oximate → —NH (CAMET) OPOB₁OPOB₂OPOB₃OPOB₄—O—DMT

oxidize → —NH—C—⟨O⟩—S⌒⌒O POB₁OPOB₂OPOB₃OPOB₄—O—DMT

beta elimination labile

1 base / 2 HOAc → HOPOB₁OPOB₂OPOB₃OPOB₄—OH

Fig. 9. CAMET linkage for DNA synthesis.

sis. The criteria for selecting protecting groups have been discussed by Reese.[28] Protecting groups should possess the following characteristics: (1) The required reagent should be commercially available or easily synthesized, (2) its introduction should occur with a high yield and the purification of the protected molecule should be relatively easy, (3) the protecting group must be stable during the synthesis and easily removable after the synthesis, and (4) its removal must cause no degradation to the product. Although other criteria may be desirable, the above are required.

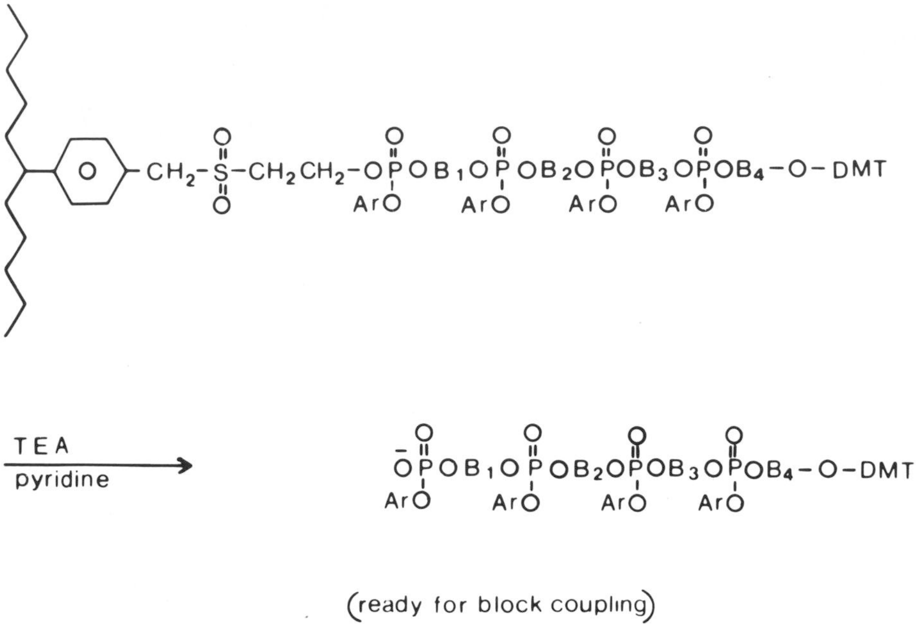

$$\text{CH}_2\text{-}\overset{\overset{O}{\|}}{\underset{\underset{O}{\|}}{S}}\text{-CH}_2\text{CH}_2\text{-O}\overset{\overset{O}{\|}}{\underset{\underset{ArO}{|}}{P}}\text{OB}_1\text{O}\overset{\overset{O}{\|}}{\underset{\underset{ArO}{|}}{P}}\text{OB}_2\text{O}\overset{\overset{O}{\|}}{\underset{\underset{ArO}{|}}{P}}\text{OB}_3\text{O}\overset{\overset{O}{\|}}{\underset{\underset{ArO}{|}}{P}}\text{OB}_4\text{-O-DMT}$$

$$\xrightarrow[\text{pyridine}]{\text{T E A}}$$

$$^-\text{O}\overset{\overset{O}{\|}}{\underset{\underset{ArO}{|}}{P}}\text{OB}_1\text{O}\overset{\overset{O}{\|}}{\underset{\underset{ArO}{|}}{P}}\text{OB}_2\text{O}\overset{\overset{O}{\|}}{\underset{\underset{ArO}{|}}{P}}\text{OB}_3\text{O}\overset{\overset{O}{\|}}{\underset{\underset{ArO}{|}}{P}}\text{OB}_4\text{-O-DMT}$$

(ready for block coupling)

Fig. 10.   BSE linkage for DNA synthesis.

A. Protection of the 5′-Hydroxyl Group of Nucleosides/Nucleotides

Protection of the 5′-hydroxyl group of 4,4-dimethoxytrityl (DMT) has become almost standard in polymer-supported oligonucleotide synthesis. Other reagents are used occasionally, such as monomethoxytrityl[29] and 9-phenylxanthen-9-yl (pixyl or PX).[30] (See Fig. 12.) All of these protecting groups serve multiple functions during the course of a synthesis. The most obvious is that they protect the 5′-hydroxyl from unwanted reactions during the preparation of the monomer coupling units and during the oligonucleotide synthesis. Since all of these groups are large lipophilic moieties, they confer a solubility in organic solvents the unprotected nucleosides would not have, thus making these intermediates much easier to purify by chromatography. The pixyl group, in addition, also confers a crystalline nature on the intermediates, and thus chromatography can sometimes be replaced by crystallization. A further useful property of these protecting groups is that they serve as indicators of the coupling reaction efficiency. The carbonium ions produced on acid-catalyzed removal of these protecting groups are intensely colored and serve as indicators of the step-by-step yield of the coupling reactions. The acid-cata-

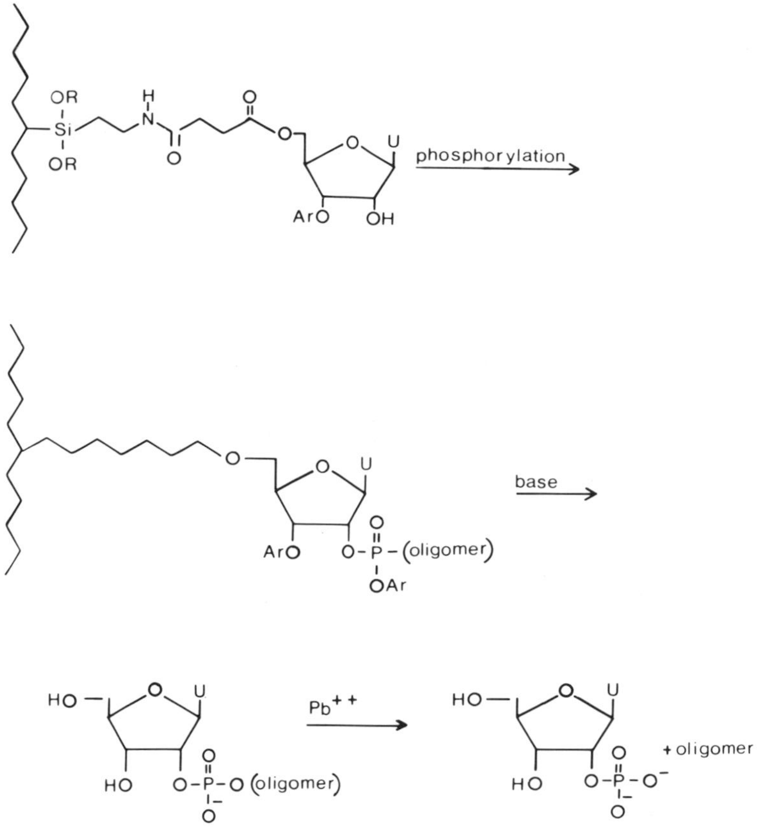

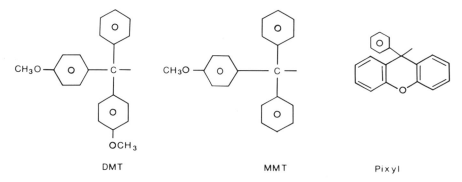

Fig. 11. Universal support for DNA synthesis.

Fig. 12. Protecting groups for the 5' OH group used in DNA synthesis.

lyzed removal of the 5′-hydroxyl-protecting group can cause cleavage of the glycosidic bond between the deoxyribose and purines. This phenomenon, termed *depurination,* will be discussed later.

B. Protection of Nucleoside Bases

The nucleosides dA, dG, and dC all contain exocyclic amino groups that have been protected by *N*-acyl groups (Fig. 13) since Khorana's original phosphodiester method.[31] These *N*-acyl groups are stable during the oligonucleotide synthesis and can be readily removed when the synthesis has been completed by ammonolysis or by oximate treatment.[32] Although these *N*-acyl groups survive during the synthesis, other portions of the molecule are not so fortunate. Thus *N*-isobutryl-dG and dT can undergo side reactions during the coupling reactions, whereas *N*-benzoyl-dA is susceptible toward depurination.

1. Protection of Deoxyguanosine

In addition to the reactive 2-amino position of dG, the oxygen at C-6 has also been shown to be susceptible to substitution and addition reactions. Thus Reese[33] reported that *N*-acyl-dG could be sulfonated in the presence of arenesulfonyl chlorides in pyridine and substituted by nitrotriazole during the coupling reaction.[34] To obviate these side reactions, Reese,[35] Rapoport[36] Jones,[37] and Pfleiderer[38] have synthesized 6-*O*-protected deoxyguanosines (Fig. 14). With longer and longer oligonucleotides being synthesized, it is imperative that these new protecting schemes be applied so that the oligonucleotide synthesized will be functional. Side reactions that occur on bases are difficult to detect in an oligonucleotide, if they occur to only a small extent (<10%). But if one in ten of the dG bases were modified, it could have a profound effect on genes synthesized from such oligonucleotides.

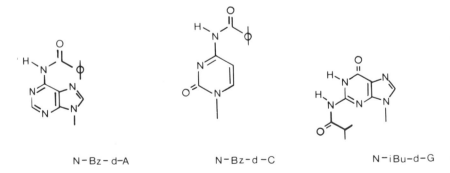

N−Bz−d−A   N−Bz−d−C   N−iBu−d−G

Fig. 13. *N*-Acyl protecting groups for the protection of nucleoside bases.

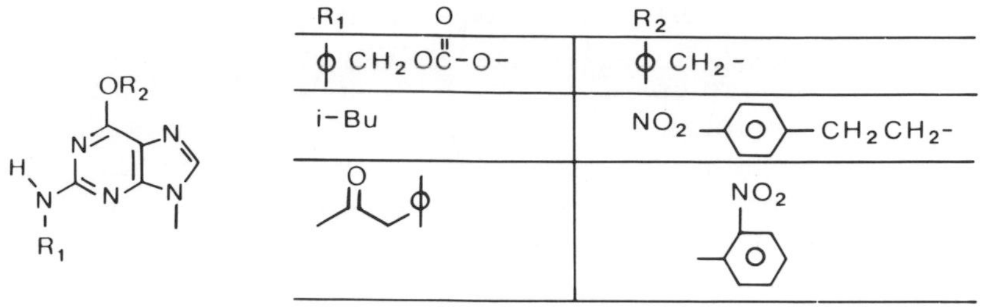

R$_1$	O	R$_2$
ϕ CH$_2$ OC-O-		ϕ CH$_2$-
i-Bu		NO$_2$ ⟨O⟩ CH$_2$CH$_2$-
		NO$_2$

Fig. 14. Groups for the protection of the 6-O of guanosine.

2. Protection of Deoxythymidine

Thymine residues are also potentially susceptible to reactions analogous to those of guanine. Sung has reported[39] that during the phosphorylation of a 5'-protected dT using 2-chlorophenylphosphoditriazolide[39], a substitution of triazole for the 4-O can occur (Fig. 15). If the triazolide is treated with ammonia before the oximate treatment, then a nucleophilic substitution can take place, forming a cytidine derivative. Reese has recently proposed that the oxygen at C-4 of thymidine be protected as a phenyl derivative to avoid such possible side reactions[35]. To avoid this side reaction during the phosphorylation and coupling reactions, most laboratories have shifted to the van Boom[40] procedure, which avoids using the triazolide for the preparation of phosphotriester and diester intermediates. It is a fortunate circumstance that many of these side reactions are reversed during the oximate deprotection step[34], but it is probably wise to avoid these side reactions if possible and not attempt to correct them after the fact.

3. Protection of Deoxyadenosine

The cleavage of the glycosidic bond (depurination) in deoxyadenosine[41] and N-benzoyldeoxyadenosine[42] are well-studied problems. The rate of

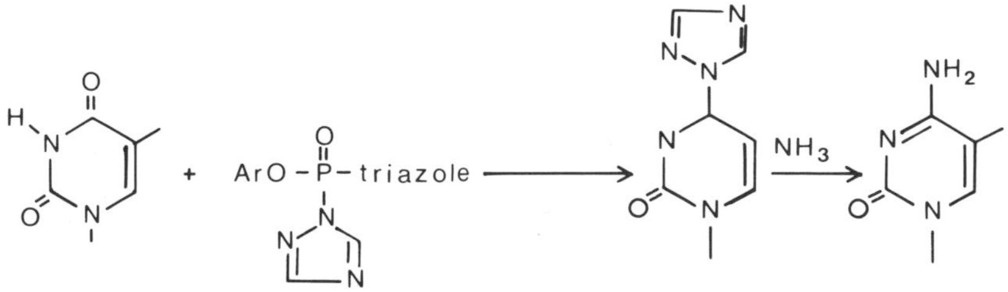

Fig. 15. Reactivity of the 4-O of thymidine.

depurination for *N*-benzoyl-dA is about six times faster than that of dA itself. An explanation for this is that dA is protonated preferentially at N-1 while *N*-bz-dA is protonated preferentially at N-7 (Fig. 16). There are three different approaches to reducing the amount of depurination that occurs during a synthesis: (1) Select a protic acid for the removal of the DMT group that causes a diminished amount of depurination. (2) Change the *N*-bz-protecting group to a different group that will convey an added stability to the glycosidic bond. (3) Change from a protic acid to a nonprotic acid.

a. Selection of a Protic Acid. The synergistic relationship between the acid chosen as the detritylation reagent and the amount of depurination has been much studied.[43] There are two types of acids that have been studied, protic and nonprotic. The protic acids are benzenesulfonic acid, dichloroacetic acid (DCA), trichloroacetic (TCA), trifluoroacetic acid, and others in numerous solvent systems. The main nonprotic acid that has been studied is zinc bromide.

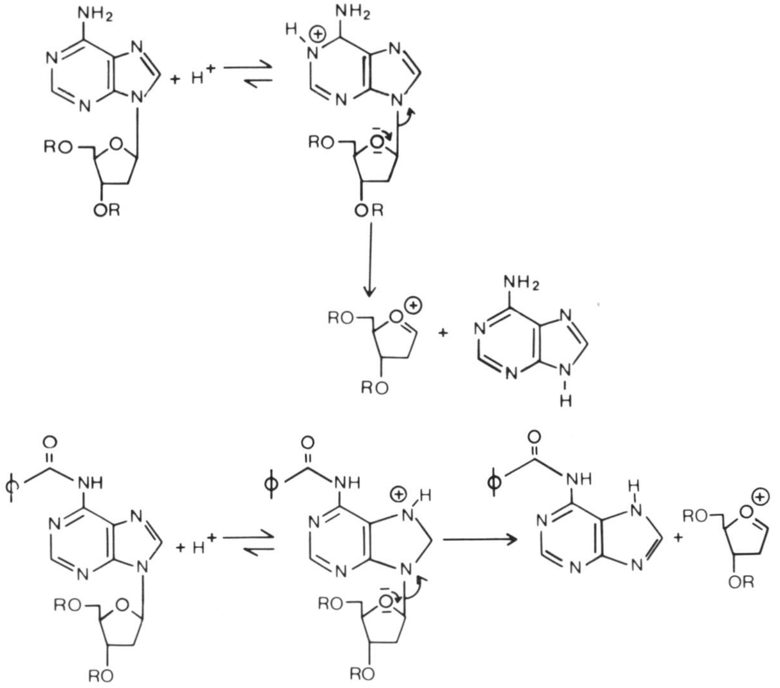

Fig. 16. Mechanisms for the depurination of deoxyadenosines.

The two acids used for most detritylation reactions in polymer-supported synthesis are DCA and TCA. Because TCA is the stronger acid, it causes more depurination in a given amount of time than DCA does. Froehler and Matteucci[43] have recently studied the rate of depurination for N-bz-dA connected to a silica support using 2% DCA in dichloromethane. Their results indicate that the half-life of the N-bz-dA connected to a silica support is only 2 hr. We have also studied the rates of detritylation and depurination on polystyrene supports. Some of our unpublished data are shown in Tables II and III.

Both the detritylation and depurination reactions are kinetically first order. Ten times the rate of detritylation was chosen because the detritylation reaction would be 99% completed in this amount of time.

As can be seen from the above data, BSA (in dichloromethane) is such a potent detritylation reagent that it would be difficult to use it because 10 times the half-life for detritylation is only 4.5 sec. Both TCA and DCA are useful as rapid and controllable detritylation reagents.

b. *Selection of a Nonprotic Acid.* A second method for decreasing the rates of depurination of N-bz-dA is to change from a protic to a nonprotic acid. Several studies have shown that the depurination problem can be significantly reduced or eliminated by switching to zinc bromide for the detritylation step.[43,45] Why then has there been no general acceptance of zinc bromide as the method of choice for detritylation. There are several complicating factors that have prevented zinc bromide from being selected as the reagent of choice. We believe that the most important of these is the fact that the rate of detritylation decreases as the length of the oligomer increases. Thus, while 5 min might suffice for the detritylation step at the beginning, by the time the oligomer length has reached 20-mer the detritylation has slowed down significantly[12]. While in the beginning of a synthesis the detritylation reaction goes to completion as the oligomer lengthens, the detritylation reaction does not quite go to completion.

TABLE II

Rates of Detritylation of DMT-bz-dA-T-polystyrene

Acid	Concentration (%)	Solvent	$10 \times t_{1/2}$ (min)
BSA	3	CH_2Cl_2/DMF (9 : 1)	0.70
BSA	0.1	CH_2Cl_2	0.036
TCA	1	CH_2Cl_2	0.32
DCA	5	CH_2Cl_2	1.35

TABLE III

Rates of Depurination of DMT-bz-dA-T-polystyrene

Acid	Concentration (%)	Solvent	$t_{1/2}$ depurination (min)	Depurination (%)/ detritylation cycle[a]
BSA	3	CH_2Cl_2/DMF (9:1)	0.0075	0.93
BSA	0.1	CH_2Cl_2	0.0047	0.076
TCA	1	CH_2Cl_2	0.021	0.15
DCA	5	CH_2Cl_2	0.0017	0.23

[a] % Depurination/detritylation cycle = $10 \times t_{1/2}$ detritylation $\times$ 100 $t_{1/2}$ depurination.

Thus, the desired oligomer of length (n) becomes contaminated with oligomers of lengths ($n - 1, n - 2, \ldots$). It is for this reason that zinc bromide has not quite lived up to its expectations as an "ideal" detritylation reagent.

c. *Selection of the Amino Protecting Groups.* The consequence of depurination is eventual fragmentation of the oligomer (during the ammonia or oximate treatment), and this problem becomes more serious for the synthesis of longer oligomers. To diminish these depurination reactions Pfleiderer,[6] Hata[47,48] and Matteucci[44] have proposed the substitution of

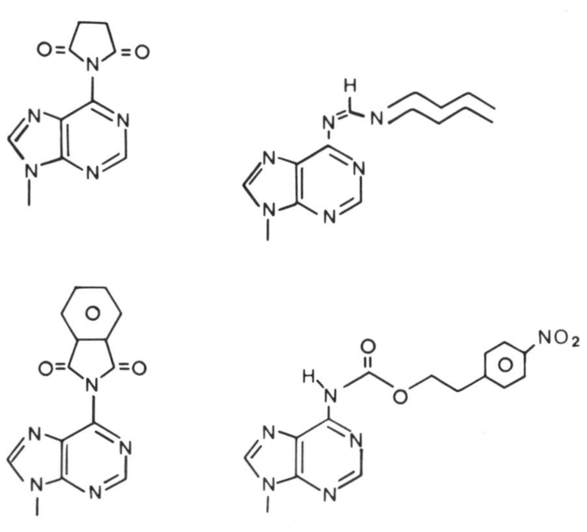

Fig. 17. Protecting groups for deoxyadenosine

the N-benzoyl group by other derivatives (Fig. 17). It is difficult to compare the results of the different protecting groups for the exocyclic amino group of dA, but it would appear that the order of stability is N-6-di-n-butylformamidine > N-6-succinyl > N-6-phthaloyl > N-6-benzoyl, while the relative rates for depurination are ~1 : 4 : 6 : 18. That is, the formamidine-dA is 18 times more stable toward depurinatiaon than benzoyl-dA.

C. Protection of the Internucleotide Phosphate

In the late 1960s Reese proposed the use of phenyl[49] and other aryl (o-chlorophenyl, p-chlorophenyl, and o-fluorophenyl) groups[50] for the protection of phosphate linkages during the phosphotriester synthesis of an oligonucleotide. The removal of these phosphate-protecting groups by alkaline hydrolysis was accompanied by an unacceptable amount of internucleotide cleavage (about 3% internucleotidic cleavage per phosphotriester group)[51]. In 1978 Reese reported the solution to this problem and proposed the use of the conjugate base of *syn*-4-nitrobenzaldoxime or *syn*-2-pyridine-2-carboxaldoxime[52]. The 2-chlorophenyl group[53] has become the standard protecting group for use in phosphotriester synthesis. It is completely stable during the synthesis and readily cleaved by oximate[52] with little or no internucleotide cleavage. The p-nitrophenylethyl group has been proposed by Pfleiderer as a replacement for the 2-chlorophenyl group, but no definitive study has yet shown that is either necessary or desirable[54].

V. SOLID-PHASE OLIGONUCLEOTIDE SYNTHESIS

The solid-phase synthesis of an oligonucleotide is accomplished by a cycle of reactions that are repeated until the final oligonucleotide has been assembled on the solid support. The oligonucleotide is then cleaved from the support and purified. This cycle of reactions includes detritylation, coupling, and capping.

1. Detritylation. The nucleoside support is always sold with a protecting group connected at the 5′ position of the nucleoside. Before the synthesis can commence this group, usually the dimethoxytrityl group, has to be removed. This is accomplished with an acidic reagent, as described above.

2. Coupling. After the 5′-protecting group has been removed and the solid support washed and made anhydrous, the incoming DMT-nucleotide is then coupled to the 5′ position of the unit on the solid support, with the

aid of a coupling reagent. This DMT–nucleotide is usually a mononucleotide but can be a di- or trinucleotide.

3. Capping. If the coupling reaction does not occur with a 100% yield, then there may be some uncoupled 5'-hydroxyl nucleotide remaining. This remaining 5'-OH is usually acetylated (capped) so that this oligomer will not be available for further coupling reactions. The cycle of reactions is repeated until the final oligomer is assembled.

A. Detritylation

Many different reagents are still being used for the removal of the dimethoxytrityl group. A representative sample of these reagents is shown in Table IV.[55–57] With the new protecting groups introduced for the protection of dA derivatives the problem of depurination may no longer be a problem. When these new methods are adopted, the primary concern will be for the completeness of the detritylation reaction.

B. Coupling Reactions

The fact that so many papers have been published on the phosphotriester method implies that the method required many improvements before it was capable of yielding oligomers greater than 20 bases long when the starting units were monomers. Attempting to compare coupling reactions from two different papers requires a careful review of the pertinent variables. Some of these variables might be (1) the chemical nature of the solid support—CPG, polystyrene, silica gel, or other; (2) the physical nature of the solid support—pore size (CPG), cross-linking(polystyrene), and particle size; (3) loading of the support; (4) the linkage between the

TABLE IV

Comparison of Various Reagents on Various Solid Supports

Acid	Concentration	Solvent[a]	Time (min)	Temp.	Ref.
ZnBr	1 M	DCM/IPA (85 : 15)	5–10	RT	17
BSA	3%	DCM/DMF (9 : 1)	3.5	5	55
DCA	3%	DCE	0.7–1.25	RT	56
ZnBr	(saturated)	DCM/IPA (7 : 3)	5	RT	18
TCA	10%	DCE	3	RT	57
TFA	2%	DCM/ACN (3 : 7)	5–7	RT	24

[a] DCM, Dichloromethane; DCE, 1,2-dichloroethane; ACN, acetonitrile; RT, room temperature.

first nucleoside and the solid support; (5) solvent; (6) concentration of reactants; and (7) temperature.

There are currently three different methods of forming the oligonucleotide linkage using phosphotriester methods. These are (1) the arenesulfonyl chloride–methylimidazole method of Efimov[58]; (2) MSNT plus methylimidazole[9]; and (3) the hydroxybenzotriazole method of van Boom[60].

Methods 1 and 2 can be accomplished on various solid supports in various solvents and the formation of the internucleotide bond requires about 10 to 15 min. Some side reactions are known to occur under these coupling conditions. These are sulfonation of the 5'-OH, sulfonation of the 6-O of dG, and base modifications of guanine and thymine.

A method proposed to avoid these side reactions is that of van Boom. In this method the nucleotide is formed *in situ* from the nucleoside and the bifunctional phosphorylating reagent[1]. (See Fig. 18.) This method is amenable to manual or automatic synthesizers and the coupling rates are fast (5–15 min). This method appears to be faster than the phosphotriester methods in which the coupling reagents are derived from sulfonic acids. The HOBT method should also be less expensive because only nucleosides are required.

The formation of the internucleotidic bond is, of course, temperature dependent. A study of the coupling reaction at 20°C and 60°C by Patel[55] showed that the reaction rate doubled for every 10° increase. The side reactions increased at about the same rate and a high-pressure liquid chromatography (HPLC) analysis showed that the quality of the synthesized oligonucleotide was comparable when synthesized at both temperatures. A recent kinetic study by Ikuta[61] showed that the reaction rates decrease as the size of the coupling unit increases from a monomer to dimer to trimer. Ikuta's data, while much more precise than that of Patel, showed similar rate changes with increasing temperature.

Two very rapid methods of forming internucleotidic phosphate bonds by the phosphotriester have been published by Froehler and Matteucci[62] and Efimov *et al.*[63] The first method[62] uses 1-methyl-2-(2-hydroxyphenyl)-imidazole as a catalytic phosphate-protecting group. The coupling rates for the ortho derivative are about 5–10 times those of the para derivative and a similar amount faster than the *p*-chlorophenyl derivative. Based on the mechanisms proposed earlier for this type of reaction, we show a possible mechanism for this reaction below (Fig. 19). Froehler and Matteuci synthesized a pentadecathymidylic acid in high yield. We shall have to see if this method can be made general for all nucleotides. The method of Efimov utilizes 4-substituted derivatives of pyridine *N*-oxide as catalysts in the condensation reaction. When the coupling reaction is carried

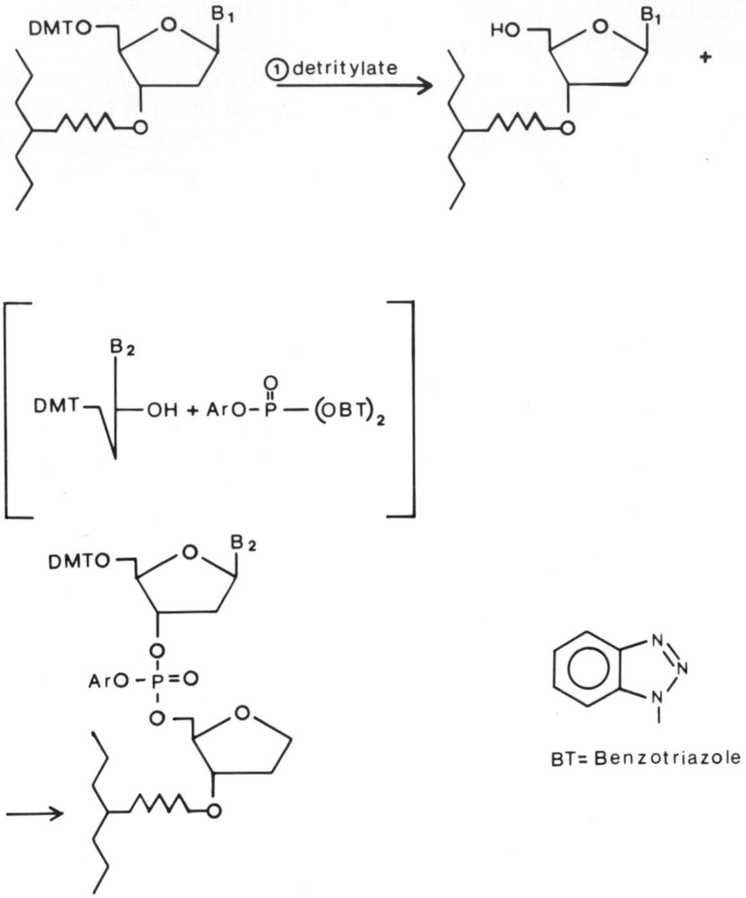

Fig. 18. Hydroxybenzotriazole method for *in situ* phosphorylation.

out with 4-ethoxypyridine *N*-oxide and MSCl in dichloromethane, the reaction is completed in less than 1 min. From this preliminary publication it would appear that this method is general for all bases.

C. Capping

The term *capping* is used to designate a reaction to block any unreacted 5'-OH groups left after the coupling reaction (Fig. 20). This reaction is usually accomplished by mixing a solution of acetic anhydride in an inert

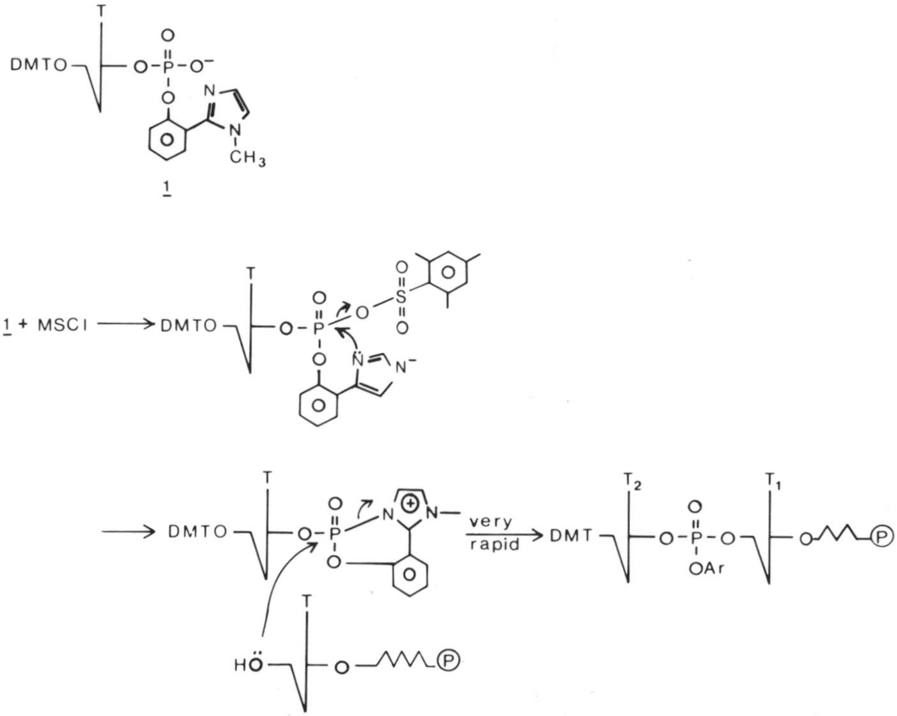

Fig. 19. Rapid phosophotriester coupling via intramolecular catalysis.

solvent, such as THF or acetonitrile, with a solution of DMAP in an inert solvent. On automatic synthesizers these two solutions are usually passed through the solid support simultaneously. These solutions must be stored separately or a very dark solution develops in less than 1 day. This capping step, which takes only 2 min or less,[63] has generated strong differences of opinion. Thus, in the recent book *Oligonucleotide Synthesis A Practical Approach,* Atkinson and Smith[64] state that the capping step "ensures that the subsequent reactions proceed only by propagating chains of the desired sequence." Another opinion is stated by Sproat and Gait in the same book. They state, "In our experience a subsequent 'capping' step . . . has not been found necessary."[9]

Our experiences indicate that the view of Atkinson and Smith is both the safer and wiser one. When we have run coupling reactions without capping, the apparent "yields" stay constant and close to 100%. Thus it would appear that the capping step is indeed removing unreacted 5'-OH nucleotides from the reaction cycle.

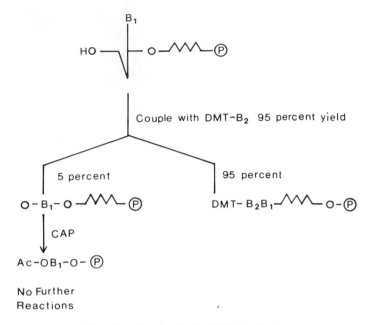

Fig. 20. Capping during DNA synthesis.

D. Standard Coupling Protocol

An Automatic machine with a continuous flow rate of 5–10 ml/min is used. The solutions are passed through a CPG support. Table V outlines the steps in the coupling procedure.

TABLE V
Standard Coupling Procedure

Reagent	Time (min)	Repeats	Notes
DCM	0.1	3	
3% DCA in DCM	1.5–2.0	2	v/v
5% TEA in CH_3CN	0.1	2	v/v
Acetonitrile	0.1	3	Dried over CaH_2
Nitrogen	0.5	1	
Nucleotide, MSCl, MeIm in DCM	15	1	Nucleotide (0.1 M), 20 molar excess; MSCl, 60 molar excess; MeIm, 60 molar excess
DCM	0.1	2	
Ac_2O, DMAP	0.3	1	20% Ac_2O in CH_3CN; 0.5 M DMAP in CH_3CN

VI. POSTCOUPLING REACTIONS

A. Removal of the Phosphate-Protecting Groups and Cleavage of the Oligonucleotide from the Polymer Support

The usual procedure for removing all protecting groups (except the DMT) from an oligonucleotide has been to first treat the polymer-supported oligomer with an oximate solution at room temperature. This removes the phosphate-protecting groups and also cleaves the succinate linkage between the 3′-nucleoside and the polymer support. The solution is then usually filtered, evaporated, and treated with concentrated ammonia for ~5 hr at ~60°C. Patel *et al.* have reported that the same oximate solution recommended by Reese, when used at 70°C, removes the chlorophenyl groups, the succinyl group, and all base-protecting groups.[65] HPLC and sequence analysis showed that the oligonucleotides were identical, when this shorter one-step deprotection method was used, to oligonucleotides produced by the standard method.

B. Removal of the *N*-Acyl–Protecting Groups

The *N*-acyl–protecting groups currently used are almost identical to those used by Khorana more than 20 years ago and their removal is usually accomplished with concentrated ammonia or oximate.[66] If an oximate treatment is not used prior to the ammonia treatment, then several side reactions that may have occurred—triazolation of dT, nitrotriazolation of dG, and sulfonation of dG—will not have been reversed and can lead to replacement by −NH groups. In addition van Boom reported that if the 5′-OH-protecting group were removed before the ammonia treatment, then unwanted side reactions could occur[66] (Fig. 21). It is not clear if these reactions will occur after the *o*-chlorophenyl group has been

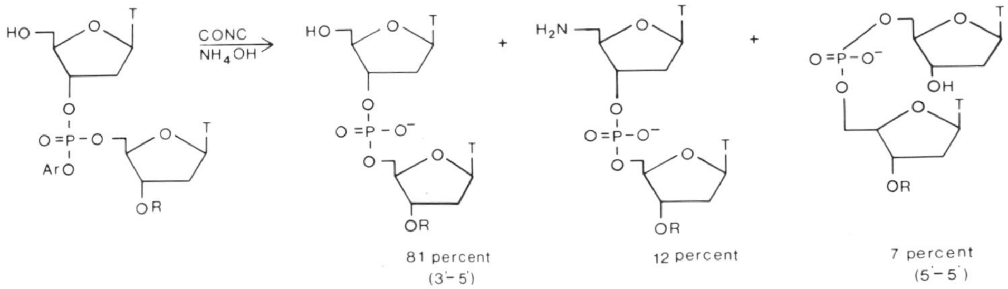

Fig. 21. Side reactions during ammonia deprotection.

removed by oximate. But until it has been shown that these side reactions do not occur upon treatment of a unprotected phosphate intermediate with ammonia, the conservative approach would be to first remove the N-acyl groups with ammonia or oximate and then remove the DMT group.

VII. NONSTANDARD COUPLING UNITS

A. Nonstandard Bases

Now that the synthesis of oligomers has become an absolutely standard procedure, some chemists are directing their attention to the synthesis of oligomers in which one of the bases is not the standard dA, dG, dT, or dC but either a modified base or a completely synthetic base. Although oligomers with base substitutions will first be subjected to studies involving their physical properties, they will eventually be used in fundamental biological studies. Thus, oligomers have recently been synthesized (Fig. 22) with O-6-methylguanine (1)[67] and 5-methylcytosine(2).[68] The production of O-6-methylguanine in DNA is thought to initiate mutagenic events. DNA that possesses a 5-methylcytosine-dG sequence is associated with decreased transcriptional levels and possibly gene inactivation.[68] Other molecules inserted into oligonucleotides have been 1,2-dideoxy-D-ribo-furanose (3),[69] 1,2-dideoxy-1-phenyl-β-D-ribofuranose (4),[69] 7-deaza-adenine (5),[70] 2-amino-2'-deoxyadenosine (6),[71] and 7-deaza-2-deoxy-guanosine (7).[72] These structures are shown in Fig. 22.

Most of the syntheses of oligomers containing nonstandard bases have been accomplished by the phosphotriester method. The primary reason for this is that the nucleotides prepared from these nonstandard bases have to be prepared and purified by the chemists who are going to do the synthesis. It is considerably easier to purify the DMT–phosphodiester nucleotides than to purify DMT–phosphoramidite nucleotides.

When the synthesis of an oligomer with a nonstandard base is attempted, the assumption is tacitly made that the nonstandard base will have properties similar to its most closely related base. As an example of problems than can occur in such a synthesis, we shall examine the synthesis of an oligonucleotide containing O-6-methylguanine.[73] The hexamer [CGC(O-6-Me)GCG] was synthesized by the phosphotriester method. This protecting group prevents unwanted side reactions that commonly occur on dG. The isobutyryl group was used for the 2-amino group of both the dG and the (O-6-Me)dG. The removal of the iBu group from the 2-N-iBu(O-6-Me)dG proved to be more difficult than expected. While this group is quantatively removed in 5 hr at 65°C in concentrated ammonia

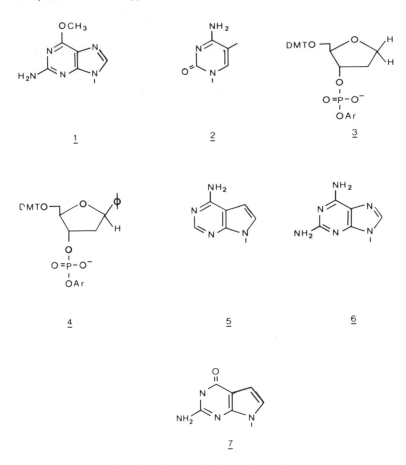

Fig. 22. Nonstandard coupling units.

from dG, the O-6-Me derivative required 3 days for the similar deprotection. When the synthesis of the oligomer had been completed and all of the protecting groups were removed, the product was subjected to a final HPLC purification step. Instead of the expected single peak, there were two resolved peaks. These were collected separately and subjected to degradation with venom phosphodiesterase and alkaline phosphatase. One of the oligomers contained the correct ratios of dC : dG : d(O-6-Me)G while the other oligomer contained the bases dC, dG, and 2-amino-dA. The 2-amino-dA must have arisen during the overly long treatment with ammonia required for the removal of the iBu group from O-6-Me-dG. A final experiment showed that this latter conversion did in fact occur slowly during the ammonia treatment (Fig. 23).

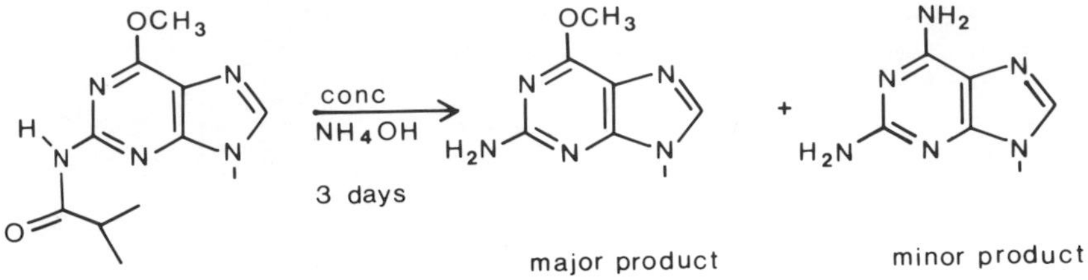

Fig. 23. The formation of 2,6-diaminopurine during the deprotection of 6-O-methyldeoxy guanosine.

The fact that the synthesis of standard oligonucleotides goes so smoothly is not because the syntheses are not frought with problems but rather that these problems have been solved by the combined efforts of hundreds of chemists over the past 30 years. The synthesis of oligomers containing nonstandard bases may require the solving of problems similar to those encountered above.

B. Phosphorothioates

A phosphorothioate nucleotide differs from the usual nucleotides in having a sulfur replacing one of the oxygens on phosphorus. The replacement of sulfur for oxygen on the phosphodiester converts the phosphorus to a chiral center. Oligonucleotides containing these phosphorothioates have been synthesized by a modified hydroxybenzotriazole phosphotriester approach by van Boom[74] (Fig. 24). The chirality of the phos-

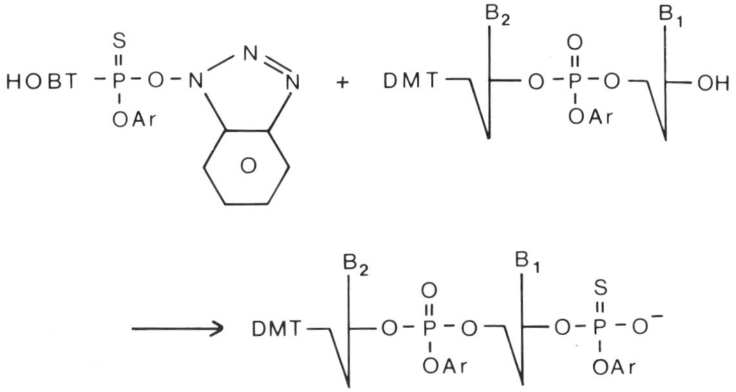

Fig. 24. Synthesis of phosphorothioates.

phorothioate was demonstrated by nuclear magnetic resonance (NMR), by chromatography, and finally by digestion by snake venom phosphodiesterase. In this method it is possible to couple either monomers or dimers. The dimers, which are diastereomeric, can be separated into their individual diastereomers by chromatography. Enzymatic treatment of the individual dimers can identify which one can serve as a substrate for the enzyme. Oligomers can then be synthesized with any specific chirality at each phosphorothioate position.

C. Methyl Phosphonates

Methyl phosphonates are nucleotide analogs in which a phosphorus–oxygen bond has been replaced by a phosphorus–methyl bond (Fig. 25). These nonionic nucleic acid analogs are capable of being incorporated into mammalian and bacterial cells and yet are resistant to nucleases.[75] Oligomers containing methyl phosphate linkages can bind to their complementary sequences and then inhibit functions of the nucleic acid within the cell. Oligonucleotides containing methyl phosphonate linkages have been synthesized by the polymer-supported phosphotriester method.

D. Oligonucleotides of Mixed Sequence

The uses of mixed oligonucleotides have been discussed recently by Wallace and Itakura.[76] The synthesis of such oligonucleotides was first reported in 1981 by Wallace.[77] They prepared a mixture of 13-base-long oligonucleotides representing eight of the possible coding sequences for amino acids 15–19 of rabbit β-globin (Fig. 26). To circumvent any problem that might arise from the differing coupling rates of different nucleotides, they chose to use mixtures of trimer blocks (Fig. 27). The trimer

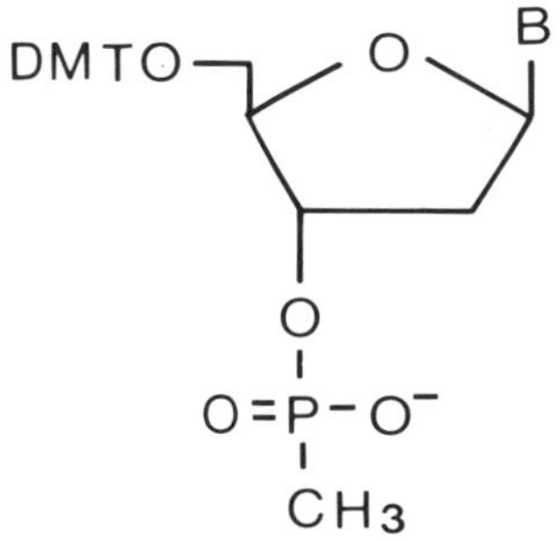

Fig. 25. Methyl phosphonate deoxynucleotide.

		15	16	17	18	19	
Amino Acid Sequence		Trp	Gly	Lys	Val	Asn	
mRNA Sequence	5'	UGG	GGC	AAG	GTG	AA	3'
Probe R β G14A	3'	ACC	CCG	TTC	CAC	TT	5'
Probe R β G14B	3'	ACC	CCG	TTC	CAT	TT	5'
Probe R β G13Mix	3'	CC	CCG	TTC	CAC	TT	5'
		A	T	T			

Fig. 26. Mixed sequences as DNA probes.

block mixtures were selected so as to have identical 3' and 5' termini with the wobble in the middle. This most conservative method was selected because the individual coupling rates of nucleotides were as yet unknown. This question was soon answered[78] and the coupling rates were indeed shown to be different. Thus the relative rates of coupling were shown to be (1 : 1 : 0.7 : 0.6) for C, T, A, G. This relatively slight difference is not usually considered critical, and when mixed oligonucleotides are required the coupling is almost always accomplished with monomers. To ensure that the correct mixture has indeed been synthesized, it is possible to verify the sequence by gel electrophoresis using a modified Maxam–Gilbert technique.[79] This method, although not quantitative, can at least indicate if the wobbles occur in the correct positions. Another method for determining the ratios of nucleotides coupled has been proposed by Fisher and Caruthers.[80] They have proposed the synthesis of nucleotides with unique triarylmethyl protecting groups. If each triarylmethyl has a spectra differing enough from the others, then a determination of the spectrum of the acid washings after detritylation should be indicative of the nucleotides coupled (Fig. 28).

An alternative approach to the use of mixed oligonucleotides as hybridization probes by insertion of deoxyinosine (Fig. 29) at ambiguous codon positions has been proposed by Ohtsuka.[81] The probes, containing five

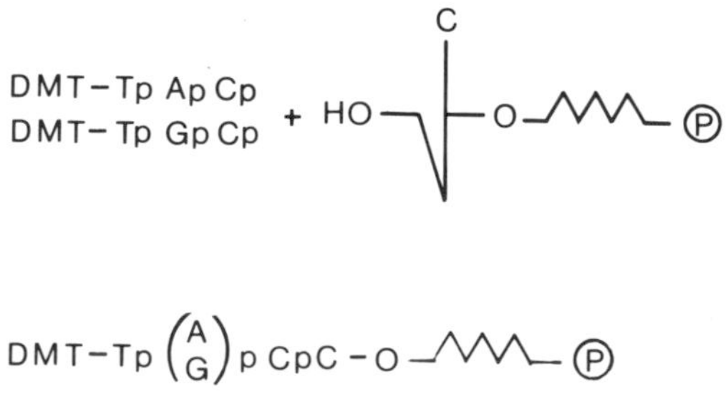

Fig. 27. First method for the preparation of mixed sequences.

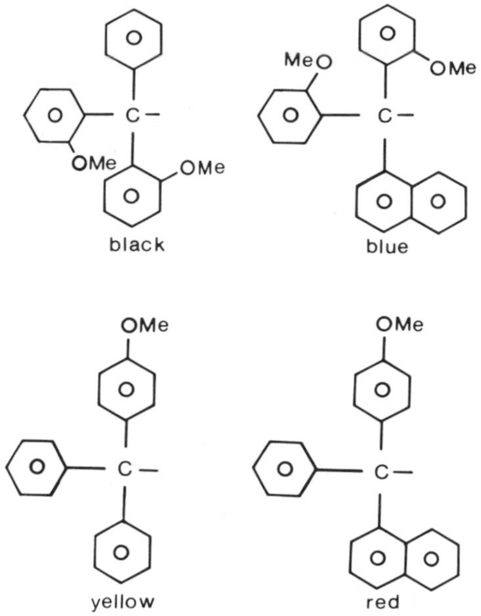

Fig. 28. Various 5'-OH protecting groups with unique spectral characteristic.

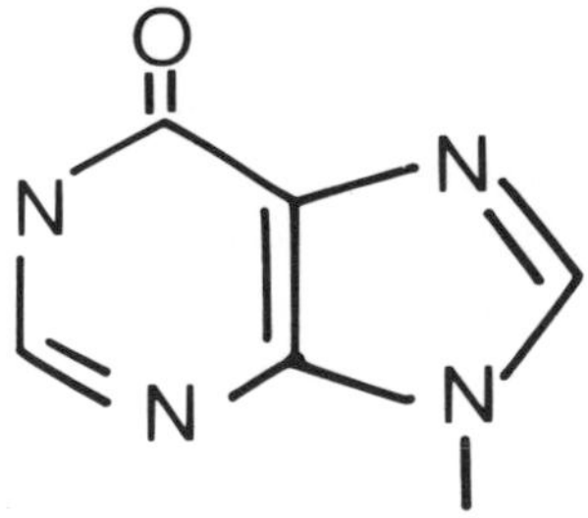

Fig. 29. Deoxyinosine, a possible universal base.

deoxyinosines, were used for screening colonies to find the desired cloned DNA sequences. The dissociation temperatures observed suggested that the insertion of deoxyinosines neither stabilized nor destabilized the DNA duplex.

VIII. AUTOMATION

The successful automation and commercialization of a synthetic process, such as the synthesis of oligodeoxyribonucleotides, depends on a number of factors: (1) a mature chemistry that is capable of producing yields in excess of 95% per step; (2) a chemistry that occurs on a solid support, so that all steps in the chemical synthesis are, in essence, reduced to washing of the growing oligomer on the solid support; (3) the availability of stable starting materials that can withstand being shipped from the manufacturer to the end user and then survive at least 1 week in solution on an instrument; (4) an instrument whose reliability approaches 100%. Until all four of these criteria were met, no automatic DNA synthesizer could be successfully commercialized. The earliest attempt (1980) to bring a DNA synthesizer to market was made by Vega. The Vega synthesizer was a scaled-down version of their peptide synthesizers. This instrument was withdrawn from the market (1984) because it could not carry out the synthesis of oligomers on a scale small enough to compete with other instruments. The next entrant into the automated oligonucleotide synthesizer market was BioLogicals (Toronto). Their instrument was not successful because their reagents were not stable.

The significance of these first-generation instruments is not that they provided a satisfactory solution over time for oligonucleotide synthesis. The real importance of these early instruments is that they laid the foundation upon which other DNA synthesizers were finally built. The same Vega instrument that was placed on the market in 1980 and given the IP-100 award as an innovative instrument in 1983, was considered obsolete in 1984.

An automated DNA synthesizer is nothing more than a device that will deliver reagents into a reaction chamber. It must be able to control the delivery with regard to both time and volume. There are three basic variables that one can discuss when considering different DNA synthesizers: (1) available chemistries, (2) software, and (3) hardware.

A. Chemistry

An "ideal" chemistry for oligonucleotide synthesis would feature the following criteria: (1) coupling yields approaching 100% with a minimum amount of excess nucleotide required to achieve this high yield; (2) starting materials stable for weeks in solution; (3) side reactions approaching 0%; (4) intermediate deprotection steps that cause no depurination during the course of the nucleotide assembly; (5) a one-step final deprotection that causes little or no internucleotidic cleavage; and (6) a relatively low cost per cycle.

The chemistry that most closely approaches the above criteria is that in which the phosphorus is protected as the diisopropylamino-O-cyanoethyl-phosphoramidite. When used in conjunction with CPG supports, this chemistry gives consistently high yields (>98–99.5%). The phosphotriester method, somewhat neglected for the past 2 years, may soon make a comeback. Matteucci[62] has been able to show that if the triester is synthesized with an intermolecular catalyst (Fig. 19), then the reaction rates for coupling are reduced from 15 min to about 1 min. Efimov has also reported coupling times of about 1 min when 4-ethoxypyridine N-oxide is used as a catalyst (63). It will take some time to see if these new chemistries can serve as a viable alternative to the amidite chemistry.

B. Software

The software in a DNA synthesizer determines how the operator interacts with the instrument and how the instrument communicates with itself. An "ideal" software would have the following features: (1) It would be menu driven, i.e., the software would lead the user through all of the required steps. (2) The programming would be easy to do and familiar to many users, i.e., it would operate like the common word processing programs. (3) It should be possible to change the chemistries and scale of synthesis with only a minimum of mechanical adjustments or programming changes. (4) The software should interact with the instrument and be able to check the status of the valves, flow rates, solvent, and reagent levels, and to insure that the instrument would stop if any mechanical failure occurred. (5) It would determine the yield of each step and shut down if a preset limit were not reached. (6) The software would aid in the changing of bottles and the purging of lines before and after a synthesis. (7) The CRT would also display information during the synthesis so that the operator will know the present status of the synthesis.

Although no instrument has software that can do all of the above, each instrument manufacturer is aware of the characteristics that make for a good software package. When some of these items are not included it is usually for cost containment reasons.

C. Hardware

The hardware in a DNA synthesizer is relatively simple. It consists of valves, tubing, and manifolds and, in some DNA synthesizers, syringes or pumps. The software controls the flow of solvents, reagents, and reactants from individually pressurized containers through a manifold and finally into and out of the reaction vessel. This flowing stream can usually be collected, monitored, or directed to a waste container. Most synthesizers can now perform mixed syntheses without the preparation of special nucleotide mixtures (Fig. 30). Although very simple in concept the DNA

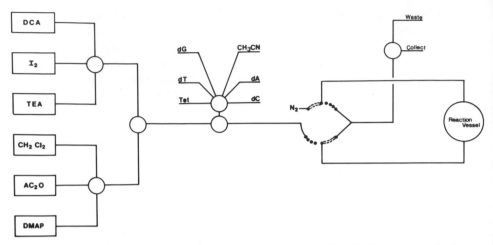

Fig. 30. Schematic of an automated DNA synthesizer, DCA, Dichloroacetic acid; I_2, iodine in water/THF/leutidine; TEA, triethyl amine in acetonitrile; CH_2Cl_2, dichloromethane; Ac_2O, acetic anhydride in THF; Tet, tetrazole; CH_3CN, acetonitrile; dG, dA, dT, dC, nucleotides.

synthesizer has required several years to overcome its early reputation for unreliability. No manufacturer uses the same valves on current instruments that were used on the first instruments. Valves, which appear to be mechanically simply and highly reliable, were found not to be so, and therefore have undergone significant changes in the past few years. Currently each DNA synthesizer uses valves that are somewhat different from those of their competitors. Whichever valves are currently in use, it is safe to say that they are much more reliable than the early valves first used.

IX. EPILOGUE

Synthetic oligonucleotides are an indispensable tool in the molecular biology laboratory and are playing a central role in the studies of molecular genetics. The solid-support synthesis of oligonucleotides by the phosphotriester method, described in this chapter, has made a significant contribution, particularly in the early developmental stages. For the last 2 years the phosphite-triester method has become more popular for the synthesis of oligonucleotides. This method has been strongly augmented by the availability of stable reagents, rapid condensation cycles, and reliable automatic machines for DNA synthesis. However if the recently published rapid coupling reactions using the phosphotriester method are shown to be of high yield and with few side reactions, we might expect a resurgence of this method. No matter which method is used for the synthesis of oligonucleotides, these syntheses have become a standard technique in all molecular biology laboratories.

ACKNOWLEDGMENT

We would like to thank Dr. Edward Felder for his careful and knowledgeable reading of this chapter. We would also like to thank Dr. Ramon Eritja for his helpful comments and suggestions through the various versions of this review.

REFERENCES

1. Merrifield, R. B. (1965). *Science* **150**, 178.
2. Letsinger, R. L., and Mahadevan, V. (1965). *J. Am. Chem. Soc.* **87**, 3526.
3. Koster, H., Pollack, F., and Cramer, F. (1974). *Liebigs Ann. Chem.*, p. 959.
4. Gait, M. J., and Sheppard, R. C. (1977). *Nucleic Acids Res.* **4**, 4391.
5. Miyoshi, K., Huang, T., and Itakura, K. (1980). *Nucleic Acids Res.* **8**, 5491.
6. Shizua, H., personal communication.
7. Itakura, K., Bahl, C. P., Katagiri, N., Michniewicz, J., Wightman, R. H., and Narang, S. A. (1973). *Can. J. Chem.* **51**, 3469.
8. Reese, C. B. (1978). *Tetrahedron* **36**, 3075.
9. Zarytova, V. F., and Knorre, D. G. (1984). *Nucleic Acids Res.* **12**, 2091.
10. Dabkowski, W., Skrzypczynski, Z., Michalski, J., Piel, N., McLaughlin, L. W., and Cramer, F. (1984). *Nucleic Acids Res.* **12**, 9123.
11. Chandrasegaran, S., Murakami, A., and Kan, L. (1984). *J. Org. Chem.* **49**, 4951.
12. Adams, S. P., Kavka, K. S., Wykes, E. J., Holder, S. B., and Galluppi, G. R. (1983). *J. Am. Chem. Soc.* **105**, 661.
13. Ito, H., Ike, Y., Ikuta, S., and Itakura, K. (1982). *Nucleic Acids Res.* **10**, 6675.
14. Koster, H. (1972). *Tetrahedron Lett.*, p. 1527.
15. Gait, M. J., Matthes, H. W. D., Singh, M., Sproat, B. S., and Titmas, R. C. (1982). *Nucleic Acids Res.* **10**, 6243.
16. Frank, R., Heikens, W., Heisterberg-Moutsis, G., and Blocker, H. (1983). *Nucleic Acids Res.* **11**, 4365.
17. Koster, H., Biernat, J., McManus, J. Wolter, A., Stumpe, A., Narang, C. K., and Sinha, N. D. (1984). *Tetrahedron* **40**, 103.
18. Kohli, V., Balland, A., Sauerwald, R., Staub, A., and Lecocq, J. P. (1982). *Nucleic Acids Res.* **10**, 7439.
19. Ito, H., Ike, Y., Ikuta, S., and Itakura, K. (1982) *Nucleic Acids Res.* **10**, 1755.
20. Crea, R., and Horn, T. (1980). *Nucleic Acids Res.* **8**, 2331.
21. Matthes, H. W. D., Zenke, W. M., Grundstrom, T., Staub, A., Wintzerith, M., and Chambon, P. (1984). *EMBO J.* **3**, 801.
22. Felder, E., Schwyzer, R., Charubala, R., Pfleiderer, W., and Schulz, B. (1984). *Tetrahedron Lett.* **25**, 3967.

23. Balgobin, N., Josephson, S., and Chattopadhyaya, J. B. (1981). *Tetrahedron Lett.* **22,** 1915.
24. Efimov, V. A., Buryakova, A. A., Reverdatto, S. V., Chakhmakhcheva, O. G., and Ovchinnikov, YA. (1983). *Nucleic Acids Res.* **11,** 8369.
25. Koster, H., and Heyns, K. (1972). *Tetrahedron Lett.*, p. 1531.
26. Gough, G. R., Brunden, M. J., and Gilham, P. T. (1983). *Tetrahedron Lett.* **24,** 5321.
27. Matteucci, M. D., and Caruthers, M. H. (1981). *J. Am. Chem. Soc.* **103,** 3185.
28. Reese, C. B. (1978). *Tetrahedron* **34,** 3143.
29. Stawinski, J. Hozumi, T., Narang, S. A., Bahl, C. P., and Wu, R. (1977). **4,** 353.
30. Chattopadhyaya, J. B., and Reese, C. B. (1978). *J. Chem. Soc., Chem. Commun.*, p. 639.
31. Agarwal, K. L., Yamazaki, A., Cashion, P. J., and Khorana, H. G. (1972). *Angew. Chem., Int. Ed. Engl.* **11,** 451.
32. Patel, T. P., Chauncey, M. A., Millican, T. A., Bose, C. C., and Eaton, M. A. W. (1984). *Nucleic Acids Res.* **12,** 6853.
33. Bridson, P. K., Narjuewucz, W. T., and Reese, C. B. (1977). *J. Chem. Soc., Chem. Commun.*, p. 791.
34. Reese, C. B., and Ubasawa, A. (1980). *Nucleic Acids Symp. Ser.* **7,** 5.
35. Reese, C. B., and Skone, P. A. (1984). *J. Chem. Soc., Perkin Trans. 1,* p. 1263.
36. Watkins, B. E., Kiely, J. S., and Rapoport, H. (1982). *J. Am. Chem. Soc.* **104,** 5702.
37. Kuzmich, S., Marky, L. A., and Jones, R. A. (1982). *Nucleic Acids Res.* **10,** 6265.
38. Trichtinger, T., Charubala, R., and Pfleiderer, W. (1983). *Tetrahedron Lett.* **24,** 711.
39. Sung, W. L. (1981). *J. Chem. Soc., Chem. Commun.*, p. 1089.
40. van der Marel, G. A., van Boeckel, C. A. A., Wille, G., and van Boom, J. H. (1981). *Tetrahedron Lett.* **22,** 3887.
41. Zoltewicz, J. A., Clark, D. F., Sharpless, T. W., and Grahe, G. (1970). *J. Am. Chem. Soc.* **92,** 1741.
42. Schaller, H., Weimann, G., Lerch, B., and Khorana, H. G. (1963). *J. Am. Chem. Soc.* **85,** 3821.
43. Tanaka, T., and Letsinger, R. L. (1982). *Nucleic Acids Res.* **10,** 3249.
44. Froehler, B. C., and Matteucci, M. D. (1983). *Nucleic Acids Res.* **11,** 8031.
45. Matteucci, M. D., and Caruthers, M. H. (1980). *Tetrahedron Lett.* **21,** 3243.
46. Himmelsbach, F., and Pfleiderer, W. (1983). *Tetrahedron Lett.* **24,** 3583.
47. Kume, A., Sekine, M., and Hata, T. (1982). *Tetrahedron Lett.* **23,** 4365.
48. Kume, A., Iwase, R., Sekine, M., and Hata, T. (1984). **12,** 8525.
49. Reese, C. B., and Saffhill, R. (1968). *Chem. Commun.*, p. 767.
50. Reese, C. B. (1970). *Colloq. Int. C.N.R.S.* **182,** 319.
51. Cusack, N. J., Reese, C. B., and van Boom, J. H. (1973). *Tetrahedron Lett.*, p. 2209.
52. Reese, C. B., Titmas, R. C., and Yau, L. (1978). *Tetrahedron Lett.*, p. 2727.
53. Reese, C. B., and Zard, L. (1981). *Nucleic Acids Res.* **9,** 4611.
54. Hillemsbach, F., and Pfleiderer, W. (1982). *Tetrahedron Lett.* **23,** 4793.
55. Patel, T. P., Millican, T. A., Bose, C. C., Titmas, R. C., Mock, G. A., and Eaton, M. A. W. (1982). *Nucleic Acids Res.* **10,** 5805.
56. Sproat, R. S., and Bannwarth, W. (1983). *Tetrahedron Lett.* **24,** 5771.
57. Marugg, J. E., McLaughlin, L. W., Piel, N., van der Marel, G. A., and van Boom, J. H. (1983). *Tetrahedron Lett.* **24,** 3989.
58. Efimov, V. A., Reverdatto, S. V., and Chakhmakhcheva, O. G. (1982). *Nucleic Acids Res.* **10,** 6675.
59. Sproat, B. S., and Gait, M. J. (1984). *In* "Oligonucleotide Synthesis: A Practical Approach" (M. J. Gait, ed.), pp. 86–115. IRL Press, Oxford.

60. Marugg, J. E., Piel, N., McLaughlin, L. W., Tromp, M., Venneman, G. H., van der Marel, G. A., and van Boom, J. H. (1984). *Nucleic Acids Res.* **12**, 8639.
61. Ikuta, S., Chattopadhyaya, R., and Dickerson, R. E. (1984). *Nucleic Acids Res.* **12**, 6511.
62. Froehler, B. C., and Matteucci, M. D. (1985). *J. Am. Chem. Soc.* **107**, 278.
63. Efimov, V. A., Chakhmakhcheva, O. G., and Ovchinnikov, Yu. A. (1985). *Nucleic Acids Res.* **13**, 3651.
64. Atkinson, T., and Smith, M. (1984). *In* "Oligonucleotide Synthesis: A Practical Approach" (M. J. Gait, ed.), pp. 35–81. IRL Press, Oxford.
65. Patel, T. P., Chauncey, M. A., Millican, T. A., and Eaton, M. A. W. (1984). *Nucleic Acids Res.* **12**, 6853.
66. de Rooij, J. F. M., Wille-Hazeleger, G., Burgers, P. M. J., and van Boom, J. H. (1979). *Nucleic Acids Res.* **6**, 2237.
67. Gaffney, B. L., Marky, L. A., and Jones, R. A. (1984). *Biochemistry* **23**, 5686; Taboury, J. A., Adam, S., Taillandier, E., Neumann, J. M., Tran-Dinh, S., Huynh-Dinh, T., Langlois d'Estaintot, B., Conti, M., and Igolen, J. (1984). *Nucleic Acids Res.* **12**, 6291.
68. Fujii, S., Wang, H. J., van der Marel, G., van Boom, J. H., and Rich, A. (1982). *Nucleic Acids Res.* **10**, 7879.
69. Millican, T. A., Mock, G. A., Chauncey, M. A., Patel, T. P., Eaton, M. A. W., Gunning, J., Cutbush, S. D., Neidle, S., and Mann, J. (1984). *Nucleic Acids Res.* **12**, 7435.
70. Ono, A., Sato, M., Ohtani, Y., and Ueda, T. (1984). *Nucleic Acids Res.* **12**, 8939.
71. Gaffney, B. L., Marky, L. A., and Jones, R. A. (1984). *Tetrahedron* **40**, 3.
72. Seela, F., and Driller, H. (1985). *Nucleic Acids Res.* **13**, 911.
73. Kuzmich, S., Marky, L. A., and Jones, R. A. (1983). *Nucleic Acids Res.* **11**, 3393.
74. Marugg, J. E., van den Berge, C., Tromp, M., van der Marel, G. A., van Zoest, W. J., and van Boom, J. H. (1983). *Nucleic Acids Res.* **12**, 9095.
75. Miller, P. S., Agris, C. H., Murakami, A., Reddy, P. M. Spitz, S. A., and Ts'o, P. O. P. (1983). *Nucleic Acids Res.* **11**, 6225.
76. Wallace, R. B., and Itakura, K. (1983). *In* "Nucleic Acid Research" (K. Mizobuchi, I. Watanabe, and J. D. Watson, eds.), pp. 227–245. Academic Press. New York.
77. Wallace, R. B., Jhonson, M. J., Hirose, T., Miyake, T., Kawashima, E. H., and Itakura, I. (1981). *Nucleic Acids Res.* **9**, 879.
78. Ike, Y., Ikuta, S., Sato, M., Huang, T., and Itakura, K. (1983). *Nucleic Acids Res.* **11**, 477.
79. Wu, R., Wu, N.-H., Hanna, Z., Georges, F., and Narang, S. (1984). *In* "Oligonucleotide Synthesis: A Practical Approach" (M. J. Gait, ed.), pp. 135–151. IRL Press, Oxford.
80. Fisher, E. F., and Caruthers, M. H. (1983). *Nucleic Acids Res.* **11**, 1589.
81. Ohtsuka, E., Matsuki, S., Ikehara, M., Takahashi, Y., and Matsubara, K. (1985). *J. Biol. Chem.* **206**, 2605.

3

DNA Synthesis for Nonchemists: The Phosphoramidite Method on Silica Supports

M. H. CARUTHERS
University of Colorado
Department of Chemistry and Biochemistry
Boulder, Colorado 80309

I. INTRODUCTION

Over the past 30 years, three procedures for synthesizing DNA have been successfully developed. Generally these methods, which are called the phosphate diester, phosphate triester, and phosphite triester approaches, are differentiated by how the internucleotide linkage of natural DNA, a phosphate diester, is initially formed. The first chemical synthesis of a dinucleotide containing a 3' → 5' internucleotide linkage identical to that occurring naturally in DNA was in 1955 by Michelson and Todd.[1] This synthesis, which was the progenitor for the phosphate triester and phosphite triester approaches, is summarized in Fig. 1. The first step involved phosphitylation of 5'-acetylthymidine (1) using a mixed anhydride (2) composed of diphenyl phosphate and benzyl phosphite. 5'-Acetylthymidine 3'-benzylphosphite (3), the product of this condensation, was oxidized to the corresponding phosphorochloridate (4) with N-chlorosuccinimide, and then condensed with 3'-O-acetylthymidine to form an intermediate phosphate triester (5). After removal of protecting groups, the product (6) was characterized as thymidylyl (3' → 5')thymi-

47

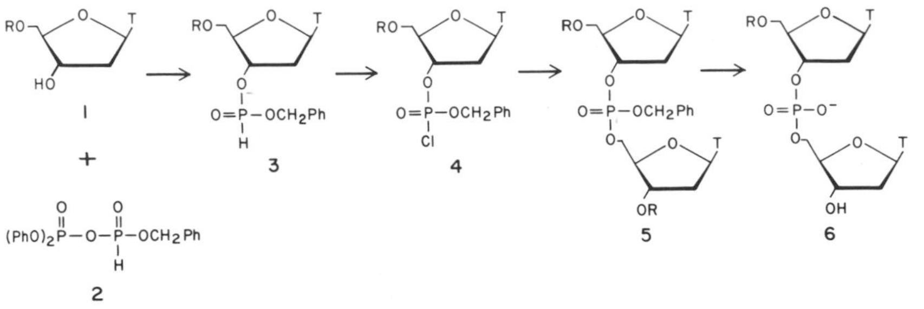

Fig. 1. The first chemical synthesis of a dinucleotide. Abbreviations: R, acetyl; T, thymine; Ph, phenyl.

dine. Despite these early achievements with activated phosphites as condensing agents and phosphate triesters as intermediate condensation products, many of the initial accomplishments in DNA synthesis were due to the phosphate diester approach as developed by Khorana in an extensive series of remarkably successful procedures.[2–4] More recently, however, the pioneering efforts of R. L. Letsinger, first using phosphate triesters[5–9] and then phosphite triesters,[10,11] have led to procedures[12–14] which overcame the earlier limitations of the work by Michelson and Todd. In this chapter, I will focus on reviewing the phosphite triester approach to oligonucleotide synthesis. Thus, I will review how nucleotide synthons as originally derived from chlorophosphines, but more recently from amino phosphines, have been used, especially in conjunction with polymeric supports, to synthesize DNA and RNA. Finally, I will summarize research involving the use of phosphite synthons to prepare polynucleotide analogs.

II. SOLUTION-PHASE PHOSPHITE TRIESTER SYNTHESIS

The chemistry of the original phosphite triester scheme for synthesizing oligonucleotides[10,11] is shown in Fig. 2. The first step involves condensation of **7**, a 5′-protected deoxynucleoside, with trichloroethylphosphodichloridite (**8a**) to form **9a**, an intermediate deoxynucleoside 3′-phosphochloridite. After a 10-min reaction time and without isolation of **9a**, a 3′-protected deoxynucleoside is added to form **10a** which is oxidized by addition of iodine and water. The product of this synthesis (**11a**) is a fully protected dinucleoside monophosphate which can be isolated free of protecting groups (**14**). Alternatively after removal of either the 5′- (com-

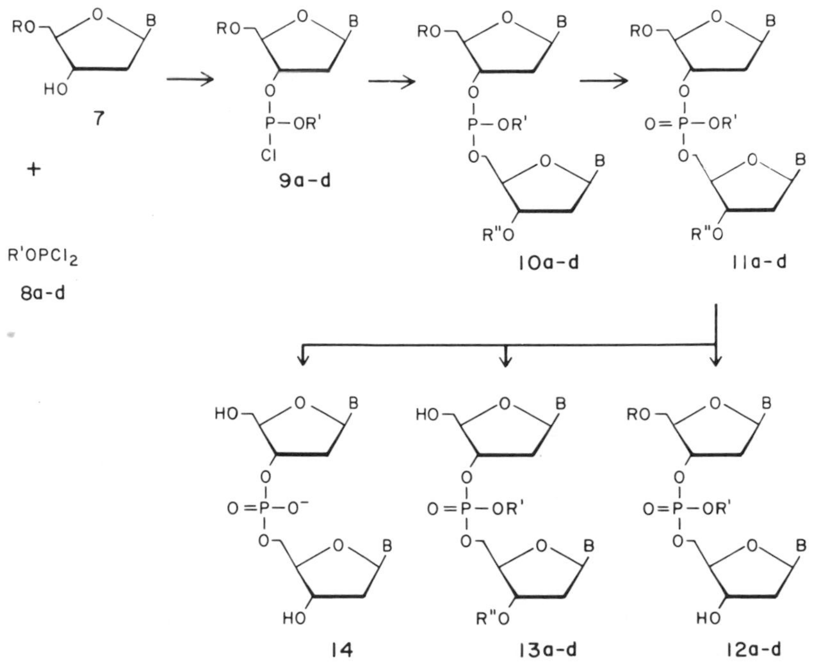

Fig. 2. The initial phosphite triester approach to DNA synthesis. Abbreviations: B, thymine or appropriately protected adenine, cytosine, or guanine; **8a–d** to **13a–d**, R′ = trichloroethyl (**a**), dimethyltrichloroethyl (**b**), *o*-chlorophenyl (**c**), and methyl (**d**). R and R″ are defined in the text. Unless specified, B as defined in this legend also covers the remaining figures.

pound **13a**) or 3′- (compound **12a**) protecting groups, the dinucleotide can be extended through the free 5′- and 3′-hydroxyls using the same chemistry to yield longer deoxyoligonucleotides.

In the initial work with this scheme,[11] it was found that **8a** could be used successfully to synthesize **11a** (R, phenoxyacetyl; R″, methoxytrityl; B, thymine) in 82% isolated yield following silica gel column chromatography. Treatment of **11a** with ammonium hydroxide yielded **13a**, which was extended to a trinucleotide, tetranucleotide, and finally a pentanucleotide (69%, 75%, and 69% yields, respectively, following silica gel chromatography) using **9a** as the synthon and intermediate deprotection steps with ammonium hydroxide to unmask the 5′-hydroxyl. For a single cycle, the overall time needed to complete the reaction was less than 1 hr and the time for workup and isolation was 5–9 hr. Thus, although this reaction sequence has more steps than either the phosphodiester or phosphotriester approaches, the speed and essentially quantitative yields of the coupling and oxidation steps provide the quickest sequence for constructing deoxyoligonucleotides.

A general problem in preparing an unsymmetrical synthon such as **9a** from a symmetrical reagent (**8a**) is that mixtures of **9a** and the symmetrical dinucleotide having a 3′ → 3′ phosphite linkage will be produced. Thus, if excess **7** is used with **8a**, considerable 3′ → 3′ dinucleotide forms, which limits the yield of **9a** and complicates the purification procedure. Conversely, with excess **8a**, it will carry over to the next step and react with the 3′-protected nucleoside to form additionally a 5′ → 5′ isomer. Initially this problem was addressed by using excess **7** and consequently converting **8a** completely to **9a** and the 3′ → 3′ isomer within 10 min at −78°C. When the 3′-protected nucleoside was then added in limiting amount, it was completely consumed. Next **11a** was separated from the 3′ → 3′ dinucleotide and the hydrolysis product of **9a** (the phosphinic acid) by silica gel chromatography. A more efficient approach involved using **8b**, which contains a bulky 2,2,2-trichloro-1,1-dimethylethyl protecting group.[15,16] This reagent was shown to react selectively with **7** to form **9b**. Unlike **8a**, only a trace of the 3′ → 3′ isomer was formed and by [31]p NMR analysis, no **8b** remained in the reaction mixture. As a test of the utility of **8b**, a deoxyhexanucleotide was synthesized by a block condensation approach. Initially **7** (R, dimethoxytrityl; B, thymine) was condensed with **8b** at −78°C to afford **9b**, which was further reacted with 3′-acetylthymidine without isolation to form **10b** and, after oxidation, **11b**. This dinucleotide was converted to **12b** and **13b** by removal of appropriate protecting groups. Compound **12b** was then reacted with **8b** followed by **13b** as the limiting reagent to yield a fully protected tetranucleotide in 93% yield. The hexanucleotide was next formed in 81% yield by first condensing **12b** with **8b** and then adding the tetranucleotide (after removal of the dimethoxytrityl group). This same reaction scheme was also used to prepare deoxyoligonucleotides containing all four mononucleotides.

Essentially an identical reaction scheme was used extensively by Ogilvie *et al.* to prepare oligoribonucleotides containing up to 16 mononucleotides.[17–22] As shown in Fig. 3, treatment of a 5′-monomethoxytritylnucleoside (**15**) with **8a** forms the 3′-phosphochloridite (**16a**) which is then condensed with 2′,3′-di-*tert*-butyldimethylsilylnucleoside (2′,3′-isopropylidene and 2′-*tert*-butyldimethylsilyl-3′-levulinyl derivatives were also used) to form the dinucleoside phosphite triester (**17a**). Further oxidation of **17a** with iodine−water forms a dinucleoside phosphotriester (**18a**). Following removal of the monomethoxytrityl group with 80% aqueous acetic acid, **18a** can be extended to form longer RNA sequences using **15, 8a**, and the same chemical cycle. These experiments demonstrated the remarkable utility of trivalent phosphites for preparing oligoribonucleotides

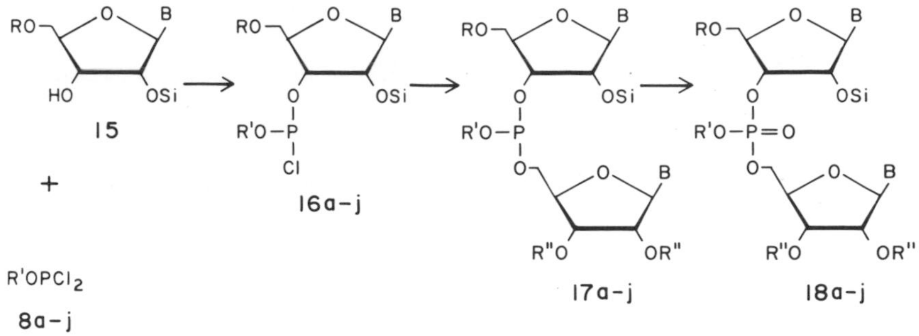

Fig. 3. Synthesis of RNA in solution using activated phosphites. Abbreviations: B, uracil or appropriately protected adenine, cytosine, or guanine; Si, *tert*-butyldimethylsilyl; R, monomethoxytrityl; **8a–j**, R′ = trichloroethyl (**a**), dimethyltrichloroethyl (**b**), *o*-chlorophenyl (**c**), methyl (**d**), tribromoethyl (**e**), β-cyanoethyl (**f**), 2-phenylethyl (**g**), 2-(*p*-nitrophenyl)ethyl (**h**), *p*-chlorophenyl (**i**), and benzyl (**j**). R″ is defined in the text.

since higher yields were maintained (70–85%) even for extended syntheses whereas with activated pentavalent phosphates, yields were generally somewhat lower (40–60%).[23] Presumably the higher yields observed with phosphites were due to less steric constraint at the 3′-hydroxyl during condensation with activated trivalent phosphorus compounds. A related result was the observation that very little 3′ → 3′ dimer (10%) was formed during activation of **15** even under forcing conditions using a one-half equivalent of **8a**. Undoubtedly the presence of the bulky 2′-*tert*-butyldimethylsilyl group reduced the yield of this side product.

Several additional phosphorus-protecting groups (**8c–j**) were investigated for synthesizing deoxyoligonucleotides and oligoribonucleotides via the reaction schemes shown in Figs. 2 and 3. When **8c** was used to form a trinucleotide from **11c**, low yields (23%) and numerous side products were observed.[11] These results were attributed to an unstable *o*-chlorophenyl phosphate triester (**11c**) toward **8c** which led to degradation. The phosphodichloridite containing a methyl protecting group (**8d**) was found to be applicable to oligoribonucleotide synthesis[24] but not especially attractive for solution-phase deoxyoligonucleotide synthesis[16] primarily because 3′ → 3′ dimers were easily formed. As observed with **8a**, the production of this side product was limited during RNA synthesis but was present in the deoxy series, even when excess **7** was employed.[25] A survey of several additional common phosphate-protecting groups under identical conditions[26] has shown that **8a–i** are all compatible with the reaction scheme shown in Fig. 3. Yields of **18a–i** generally were around 70% with these phosphites, whereas with **8j**, the yield of **11j** was only 20% which suggests that the benzylphosphodichloridite was unstable to the reaction conditions normally used for preparing dinucleotides. Unfortunately a thorough comparison by converting these dinucleotides to trinucleotides was not completed. Thus, unlike earlier reports with the o-chlorophenyl phosphodichloridite,[11] trichloroethyl,[11] 1,1-dimethyl-2,2,2-trichloroethyl,[15,16] and methyl,[24] no data was presented that addresses the stability of the phosphate triesters carrying these protecting groups to reagents **8e–j**. For example, it is curious that **8c** can be unsatisfactory[11] whereas **8i** can be considered acceptable.[26]

III. SOLID-PHASE PHOSPHITE TRIESTER SYNTHESIS

A. Introduction

For some time now, an attractive alternative methodology for synthesizing oligonucleotides has been to immobilize the growing segment on an insoluble carrier during the synthesis process. Because oligonucleotide synthesis uses the same reagents as part of a cyclical process, this concept offers several advantages over conventional solution chemistry. These include the reduction of intermediate purification steps to simple washing or filtration procedures rather than the extensive chromatographic methods required for intermediate solution-phase synthesis and an opportunity to automate the process. Moreover the simplicity of the separation procedures allows the reactants to be used in large excess, which maximizes the yield of the desired product. Thus, this technique is especially useful for small-scale syntheses (most applications for synthetic oligonucleotides in molecular biology and biochemistry require only milligram amounts of material) where the use of excess reagents is least costly. When large quantities of oligonucleotides are needed (hundreds of milligrams), solution-phase chemistry may be advantageous because reactants are mixed in equimolar amounts to maximize yields for the least cost.

During the past several years, both the phosphite triester and the phosphate triester approaches have been successfully adapted to the rapid synthesis of polynucleotides on a solid phase. Because the phosphite triester approach gives higher step yields (near quantitative) and fewer side products, it has become the method of choice for most synthesis applications today and will be reviewed in this section.

B. Solid-Phase Supports for the Phosphite Triester Approach

To be useful as a matrix for synthesizing oligonucleotides, a solid phase must meet certain criteria. These include the rapid diffusion of reactants, solvents, and products into and out of the matrix, a structure that is chemically inert toward all reagents needed during the synthesis cycle, and a readily accessible functional group for attaching the growing oligonucleotide to the support. In the latter case, the covalent link joining the oligonucleotide to the matrix must be stable toward synthesis reagents as well and yet hydrolyzable under conditions that do not destroy the product. As the result of an extensive search, a large number of insoluble supports have been tested. These include polydimethylacrylamide,[27,28] polyacrylmorpholides[29,30] polystyrene,[5-7,31,32] polystyrene grafted on polytetrafluoroethylene,[33] cellulose,[34] silica,[25,35-37] and porous glass.[38-40] Among these, silica and porous glass are particularly well suited since the particles do not swell or contract in various solvents and efficient mass transfer throughout the matrix is observed. Thus, despite only marginal early results which did not suggest that silica would be a suitable matrix,[37] silica-based supports were developed successfully as part of the phosphite triester approach[25,36] and more recently the phosphate triester approach as well.[41,42]

The general procedure as currently used for covalently bonding deoxynucleosides to silica is shown in Fig. 4. Compound **19**, a silica-based support derived from either silica gel or porous glass can be selected from

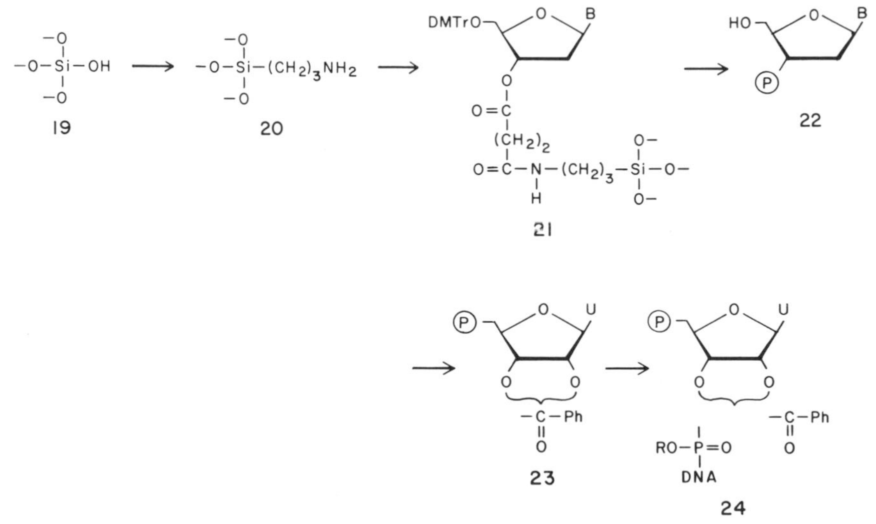

Fig. 4. Synthesis of a nucleoside covalently bonded to a silica support. Abbreviations: U, uracil; Ph, phenyl. For compounds **23** and **24**, the benzoyl group and the growing DNA segment can be attached to either the 2'- or 3'-hydroxyl. Ⓟ, as shown in compounds **22**, **23**, and **24**, is an abbreviation for the silica matrix as defined in compound **21**. Ⓟ in the remaining figures is defined in an analogous manner.

various matrices which are known to perform with efficient mass transfer in high-performance liquid chromatography (HPLC). So far the best results have been obtained from commercial samples such as Vydac TP-20,[25,36] Fractosil,[43,44] and porous glass.[38–40] Silica (19) is first treated with 3-(triethoxysilyl)propylamine in refluxing toluene to yield 20, which is then converted to 21 by one of two pathways. By treatment first with succinic anhydride followed by 5'-dimethoxytrityldeoxynucleosides and dicyclohexylcarbodiimide (DCC), a two-step procedure can be used to generate 21.[25,36] Alternatively, 5'-dimethoxytrityldeoxynucleoside can first be converted to the 3'-succinate half-ester using succinic anhydride, then this half-ester activated to p-nitrophenylsuccinate ester using DCC and p-nitrophenol, and finally the activated ester condensed with 20 to form 21.[45–47] Although the former pathway is simpler to execute, the results are more sporadic since loading is quite variable and can range from 1 to 50 μmole nucleoside/gram of silica. Presumably this is due to the nonuniform bonding of the amine to 19[48] which can eventually lead to formation of carboxylic acid anhydrides on the support during the DCC activation step. The formation of these anhydrides could effectively reduce the yield of 21. The second approach where an activated nucleoside is formed in the solution phase and then condensed with 20 to form 21 is much more easily controlled and leads to reproducible loadings. Irrespective of which method is used, excess silanol and amino groups must be chemically blocked using appropriate reagents to mask secondary reaction sites during oligonucleotide synthesis.[25] Compound 21 derivatized to contain each of the four deoxynucleosides can then be treated with acid to remove the dimethoxytrityl group and form 22, the starting material for deoxyoligonucleotide synthesis.

Recently 2'(3')-O-benzoyluridine linked 5' to silica (23) has been proposed as a universal linker for DNA synthesis.[49] The growing deoxyoligonucleotide is joined through either the 2'- or 3'-hydroxyl of uridine to the support (24). After completion of the synthesis and removal of protecting groups, the segment is freed of uridine 2'(3')-phosphate by treatment with Pb^{++}, which has been reported to hydrolyze completely the uridine 2'(3')-deoxyoligonucleotide linkage (via a cyclic phosphate intermediate). The advantage of this approach is that only one support containing uridine is needed for any deoxyoligonucleotide attachment site. This is because the first nucleoside of the segment being synthesized is carried to the support as an activated 3'-phosphate or phosphite, covalently joined to uridine, and then subsequently removed at the end of the synthesis. Conversely when 22 is used, the 3'-nucleoside of a growing segment is the nucleoside attached directly to silica. Thus, four different resins are needed to have total sequence flexibility. This matrix (23) could therefore prove to be quite useful as a universal support if further work confirms that metal ions do indeed quantitatively cleave the $3' \to 3'$ and $2' \to 3'$ linkages joining uridine to the deoxyoligonucleotide.

C. Solid-Phase Synthesis Using Chlorophosphites

The procedure shown in Fig. 5 outlines the first successful synthesis of DNA using phosphites on a solid-phase support (silica).[25,35,36] A deoxynucleoside linked to silica through the 3'-hydroxyl is condensed with excess of the appropriately protected deoxynucleoside 3'-chlorophosphite (25) to form 26 which, after capping, oxidation, and detritylation with acid, yields 27, a dinucleotide that can now be recycled repetitively using the same chemistry to yield a deoxyoligonucleotide of the desired length and sequence. Therefore the DNA synthesis strategy involves linking the growing segment to silica through a base-labile bond and extending the sequence using synthons transiently protected with acid-labile trityl ethers. Finally after the addition of an appropriate series of nucleotides, the phosphorus methyl-protecting group is removed with thiophenol and then, upon treatment with ammonium hydroxide, the deoxyoligonucleotide is hydrolyzed from the support and freed of exocyclic amino-protecting groups located on the nucleoside bases using concentrated ammonium hydroxide. Alternatively, ammonium hydroxide has been proposed as a reagent for removing the methyl and exocyclic amino-protecting groups and for hydrolyzing the deoxyoligonucleotide from the support.[50] One report, however, suggests that thiophenol is preferred and the use of ammonium hydroxide to hydrolyze the internucleotide methyl phosphate triester could lead to the formation of internucleotide phosphoramidates (to a small extent) by attack of ammonia on phosphorus rather than the methyl group.[51] Using this chemical synthesis procedure, DNA segments corresponding to modified *lac* operators[25] and the SV40

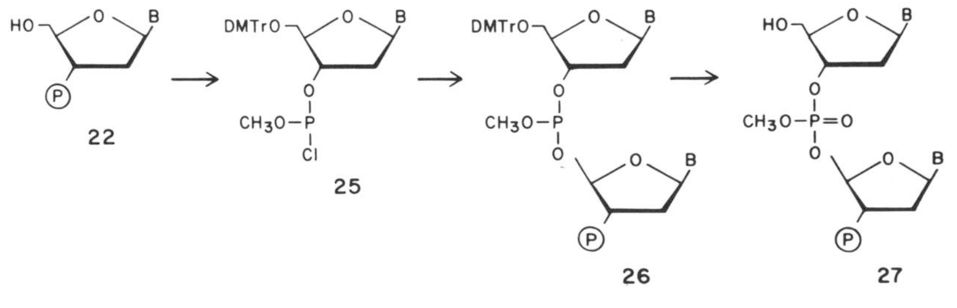

Fig. 5. Synthesis of DNA using silica supports and nucleoside 3'-chlorophosphites. Abbreviation: DMTr, dimethoxytrityl.

T-antigen binding site I[52] have been synthesized and shown to be bio-chemically active.

As can be seen by examining Fig. 5, syntheses were completed in a $3' \to 5'$ direction rather than $5' \to 3'$ and the activated chlorophosphite (25) was in the solution phase. There are several reasons for selecting this approach. (i) Synthons having a free 3'-hydroxyl and protected with a 5'-O-dimethoxytrityl ether are easier to synthesize and require fewer steps than deoxynucleosides having free 5'- and protected 3'-hydroxyl groups. (ii) Higher yields would be expected when 3'-chlorophosphite is used as a condensing agent with a free 5'-hydroxyl group. The converse situation, at least when activated 5'-phosphate monoesters[53] or diesters[6,54] are used to phosphorylate 3'-hydroxyl groups, generally leads to lower yields of internucleotide linkages. (iii) By using an excess of 25, the condensation step can be driven to essentially quantitative yield and, since the excess is in the solution phase, it can be removed simply by filtration. (iv) This excess chloridite also serves as its own desiccant since any water in the solvent will react with 25 to form an unreactive phosphinic acid which again can be removed by filtration. Recently a variation of this approach was proposed whereby 22 is first phosphitylated with methoxydich-lorophosphine (8d) and then the internucleotide bond is formed by addi-tion of 5'-O-dimethoxytrityldeoxynucleosides.[55] Clearly the most serious problem with this approach is hydrolysis of the highly reactive chlorophosphite attached to the 5'-hydroxyl of the support bound growing DNA segment (see also ii above). Even trace amounts of water in solvents will quench this chlorophosphite, lead to reduced step yields, and conse-quently severely restrict the length of DNA that can be prepared.

As expected in a repetitive synthesis where yield maximization and side product formation are important considerations, the various chemical steps, including capping, oxidation, and detritylation, have been carefully examined. Capping, which refers to irreversible blocking of growing DNA segments that fail to react with 25 (i.e., failure sequences), should be included in any polymer-supported synthesis procedure. In this way an incrementally small amount of a failure sequence accumulates during each cycle and, most importantly, these blocked segments are not ex-tended with 25 during subsequent synthesis cycles. Conversely, without capping, failure sequences are extended to a large, heterogeneous family of DNA segments which primarily lack one or two nucleotides. These compounds are difficult to remove from the product using standard purifi-cation procedures whereas the shorter failure sequences that accumulate as the result of a capping procedure are more easily removed. Initial capping reagents included phenyl isocyanate[35,36] and diethoxytriazoy-lphosphine.[25] Now, however, acylation using acetic anhydride with N,N-

dimethylaminopyridine as catalyst has been shown to be the reagent of choice[45,46] since acylation is complete in less than 1 min without generating side products. The next repetitive step involves oxidation of the phosphite to the phosphate using aqueous iodine. Due to the lability of phosphites to the acids used during detritylation, oxidation yielding acid-stable phosphates must be included in each nucleotide addition cycle. Attempts to postpone oxidation until after all condensation steps lead to low yields and uncharacterized side products.[25] Although aqueous iodine is extremely mild, fast (complete reaction within a few seconds), and does not lead to production of side products, several other oxidants such as *m*-chloroperbenzoic acid[51,56] perbenzoic acid,[57] *tert*-butyl peroxide,[57] and hydrogen peroxide[57] have been investigated. Of these, only *m*-chloroperbenzoic acid gives results comparable to aqueous iodine and then only if pyridine is present.[51] The others either react slowly (several hours for complete oxidation) or lead to the production of uncharacterized side products. The final repetitive step is acid detritylation, which can be completed using either protic or Lewis acids. Because acidic conditions can lead to extensive depurination, primarily of *N*-benzoyladenine from deoxyadenosine derivatives, a large number of acids have been examined for removing dimethoxytrityl-protecting groups.[12–14] Of these trichloroacetic acid in nitromethane–methanol[51] and dichloroacetic acid in dichloromethane[40] appear particularly attractive among protic acids. Because dichloroacetic acid is more stable toward decomposition to hydrochloric acid, it is preferred. Two groups independently introduced zinc bromide as a selective detritylating agent[58,59] because it does not cause depurination in anhydrous nitromethane, even over a 24-hr period.[25,51] For rapid syntheses on solid supports, zinc bromide in anhydrous nitromethane however is not very effective since at least 30 min is required for complete reaction.[25] Recently mixed solvents such as nitromethane–methanol (5%),[45] nitromethane–water (1%),[46] and dichloromethane–2-propanol[60] were shown to lead to rapid (less than 5 min) detritylation without depurination. When 10% methanol in chloroform was used with 0.7 M $ZnBr_2$, detritylation was complete in 1 min whereas overnight treatment led to 30% debenzoylation of *N*-benzoyladenine. In an 80% methanol–20% chloroform solution, 1 M $ZnBr_2$ led to complete debenzoylation within 6 hr.[60] Based on this data using forcing conditions, it has been concluded erroneously[12,60] that 5% methanol in nitromethane containing 0.1 M zinc bromide and a 5-min reaction time[45] should not be used for synthesis because debenzoylation of adenosine would occur. A more serious problem with zinc bromide, however, is that the rate of dimethoxytrityl removal decreases rapidly as the DNA segment becomes longer.[40,61] This observation effectively limits zinc bromide usage to the synthesis of deoxyoligonucleotides containing less than 15 deoxymononucleotides. Since depurination has not been observed for any zinc bromide detritylation solution (anhydrous or mixed with alcohols or water), perhaps the major application of this Lewis acid to polynucleotide synthesis will be in the preparation of segments which are to be used for biophysical studies where DNA free of depurination is important.[52,62–65]

An alternative strategy for stabilizing deoxyadenosine toward depurination is to change the exocyclic amino-protecting group. Of the protecting groups examined to date, the most promising appear to be the succinoyl[66] and various amidines[67–69] in that an approximately 20-fold stabilization has been achieved in each case over N-benzoyldeoxyadenosine. The introduction of these protecting groups on deoxyadenosine should improve the overall, isolated yields of DNA.

Care must also be taken during the preparation of **25**. This reagent is synthesized by reacting the appropriate 5'-dimethoxytrityldeoxynucleoside and methoxydichlorophosphine (**8d**) in anhydrous tetrahydrofuran (THF) containing collidine. After removal of collidine hydrochloride by filtration, it is important that unreacted **8d** be removed prior to use of **25** in DNA synthesis. Otherwise the methoxydichlorophosphine will rapidly phosphitylate **22** and reduce the overall yield of growing deoxyoligonucleotide. Removal of this reagent can most easily be accomplished by repeated evaporation from toluene[25] or pyridine.[51] A failure to recognize this problem may have contributed to lower than expected yields during certain early attempts to adapt dichlorophosphites to deoxyoligonucleotide synthesis on polymeric carriers.[50] Since **8d** is present even with excess nucleoside,[25] perhaps a ring adduct forms which either is in rapid equilibrium with starting materials or is destroyed during the repeated evaporation with pyridine and toluene. Repeated reconcentration from added solvent then eliminates the methyldichlorophosphite as it is freed from the unstable adduct.

D. Solid-Phase Synthesis Using Phosphoramidites

Although deoxynucleoside 3'-phosphochloridites (**25**) were used to synthesize a large number of DNA segments on silica supports, several problems were routinely encountered in their use. For example, methyldichlorophosphite which was used to synthesize **25** must be freshly prepared and stored under an inert gas atmosphere at $-20°C$ because it is unstable toward disproportionation, is easily oxidized or hydrolyzed, and is pyrophoric. Additionally the product of this synthesis (**25**) suffers from these same chemical properties and must be manipulated in solvent using syringes or a cannula under an inert gas atmosphere while the system is accessible to a vacuum line. At best **25** is stable for only a few days if stored at −20°C over an inert gas atmosphere. These reaction conditions are neither familiar to nor easily practiced by nonchemists, the primary users of synthetic DNA. Thus, although the step yield for internucleotide bonds is very high and the extent of side product formation is lower than with the phosphate triester approach, the phosphite method as outlined in Fig. 5 is generally accessible only to chemists. Consequently new synthons utilizing phosphoramidites were investigated.

Initially deoxynucleoside phosphoramidites were prepared from **25** by additional reaction with tetrazole, triazole, and 4-nitroimidazole to form **28a**, **28b**, and **28c**, respectively,[25,35,47] and then used to synthesize DNA as outlined in Fig. 6. Based on a quantitative HPLC analysis of the dinucleotide products hydrolyzed from silica supports, **28a** gave higher yields than any of the others, including the chloridite, under identical reaction conditions. Moreover, unlike the chloridites, the tetrazolides were stable for approximately 3 months when stored as anhydrous foams under an inert gas at −20°C. Compounds **28a** and **28b** have also been prepared by first converting methoxydichlorophosphine (**8d**) to the bistetrazolyl and bistriazolyl derivatives.[70] When equivalent amounts of 5'-O-dimethoxytritylthymidine followed by 3'-O-acetylthymidine were added to these heterocyclic amidites, the dinucleotide products were isolated in high yield with very little 3' → 3' dimer being produced. All these results suggest that phosphoramidites are less reactive than chlorodites in much the same

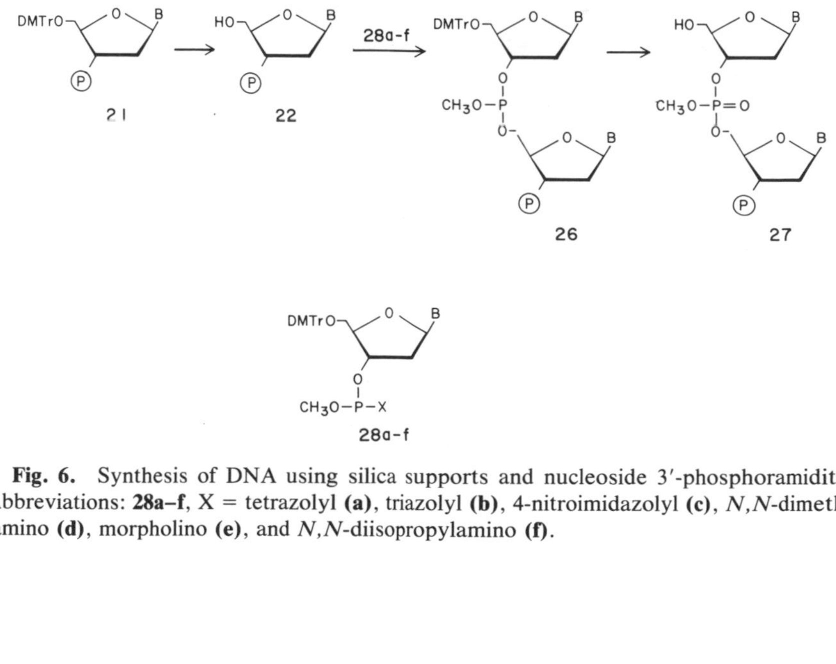

Fig. 6. Synthesis of DNA using silica supports and nucleoside 3'-phosphoramidites. Abbreviations: **28a–f**, X = tetrazolyl (**a**), triazolyl (**b**), 4-nitroimidazolyl (**c**), N,N-dimethylamino (**d**), morpholino (**e**), and N,N-diisopropylamino (**f**).

way that phosphorotriazolides and phosphorotetrazolides are less reactive than phosphorochloridates. As was also the case for the pentavalent analogs, phosphitylations were observed to be more selective and to give higher yields with the heterocyclic amines rather than the chloridites. A related approach,[71] comparable in synthesis strategy to Jayaraman and McClaugherty,[55] involves condensation of **22** with bistetrazoylmethoxyphosphine to form a support-bound nucleoside 5'-methoxytetrazolylphosphine, which subsequently reacts with excess 5'-O-dimethoxytrityldeoxynucleoside to form **26**. Although the method is claimed to give high repetitive yields, it is not a practical approach unless used in a vacuum line with rigorously anhydrous reagents and solvents. The problem of course is that the intermediate, activated phosphite is present as the limiting reagent and is attached to the silica support. Thus, any trace water contamination, which is always present, will hydrolyze the phosphite, terminate synthesis, and reduce the overall yield. A serious problem generally with these heterocyclic amidites is their lack of stability without special precautions toward oxidation and hydrolysis. They must also be prepared from highly reactive, pyrophoric reagents just prior to usage.

Phosphoramidites generally useful to the nonchemist were first introduced by Beaucage and Caruthers to control the reactivity of the active phosphite.[72] Their use in DNA synthesis on silica supports is also summarized in Fig. 6. Thus, appropriately protected deoxynucleoside 3'-N,N-dimethylaminomethylphosphoramidites **(28d)** were dissolved in acetonitrile containing tetrazole and reacted with **22** to form **26**. The synthesis was complete within a few seconds with **26** being formed in essentially quantitative yield. After capping, oxidation, and detritylation as described in the previous section, **27** was ready for repeated recyclings with additional deoxynucleoside 3'-phosphoramidites to form a DNA of the desired length. Unlike the chloridites and heterocyclic amidites, these synthons are remarkably stable toward oxidation and hydrolysis. (One step in the synthesis of **28d** is an aqueous extraction.) They can be stored indefinitely until needed as stable solids, but when activated with a weak acid, an extremely reactive phosphite forms which condenses essentially quantitatively within a few seconds with **22**. Although several weak acids were found to activate these amidites, tetrazole was the acid of choice because it is a nonhygroscopic solid that can be easily purified by sublimation. Many of the other acids examined were unacceptable, especially for routine use by nonchemists, because they easily adsorbed water which quenches the condensation step and leads to considerably lower DNA yields. The mechanism of activation is only partially understood at this time. The first step presumably involves protonation either on nitrogen or phosphorus followed by nucleophilic attack of the anion on phosphorus.

For example, early ^{31}P NMR data demonstrated that **28d** was converted to the chlorophosphite **(25)** when treated with N,N-dimethylaniline hydrochloride.[72] The chlorophosphite being the reactive species then condenses with a nucleoside to form a dinucleotide phosphite. With tetrazole, an analogous mechanism involving protonation followed by nucleophilic attack on phosphorus would lead to a highly reactive tetrazolide **(28a)** (^{31}P NMR signal at -126.9 ppm) which then condenses with a second nucleoside.[69,73] Additional evidence on the formation of an intermediate tetrazolide comes from recent work by Seliger and Gupta.[74] In their work, **25** was reacted with polystyrene or silica gel containing either ethylamino or piperazino groups to form a deoxynucleoside N,N-dialkylaminophosphoramidite attached to an insoluble support. When samples of these supports were treated with tetrazole in acetonitrile and the solutions syringed to support-bound nucleosides **(22)**, the dinucleotide was observed to form in near quantitative yield.

Several additional deoxynucleoside phosphoramidites have been investigated recently. Generally variation in structure was introduced by changing the functionality on nitrogen or the blocking group on phosphorus. Table I lists a summary of these amidites with appropriate references. The initial impetus for examining different dialkylamino groups was that N,N-dimethylaminophosphoramidite preparations varied considerably in purity and stability. These investigations led to the development of the morpholino-**(28e)** and the N,N-diisopropylamino-**(28f)** phosphoramidites by McBride and Caruthers.[73] They demonstrated that these amidites, when isolated in essentially pure form simply by an aqueous extraction of reaction mixtures, remained stable indefinitely as solids or for weeks in acetonitrile solution. Also in a manner similar to **28d**, these amidites could be activated with tetrazole and used to form dinucleotide phosphites. They also demonstrated that the morpholino derivative, unlike the N,N-dimethylamino- or N,N-diisopropylaminophosphoramidites could be purified to homogeneity by silica gel column chromatography. These results were later confirmed by Dorper and Winnacker[82] and also Adams *et al.*[40] A potentially serious problem with **28e** is its lower rate of activation with tetrazole. Thus, under conditions where complete reaction was observed with **28d** or **28f** to form **26** (1 min), **28e** had reacted only to 70% completion. Additionally upon treatment with tetrazole, the ^{31}P NMR spectrum of **28e** showed only 5% conversion to the tetrazolide whereas **28f** was 95% converted to this active phosphitylating agent.[73] Because **28e** can easily be purified to homogeneity by silica gel column chromatography and is stable to air oxidation and hydrolysis, several investigations have focused on identifying acids that can be used to activate the morpholidite more completely. These studies have shown that

TABLE I
Summary of Deoxynucleoside Phosphoramidites

R′	R″	R′″	−NR″R′″	References
1. Methyl	Methyl	Methyl		72, 77
2. Methyl	Isopropyl	Isopropyl		73, 40
3. Methyl			Morpholino	73,74, 76, 78
4. Methyl	Ethyl	Ethyl		40
5. Methyl	Isopropyl			40
6. Methyl				73,75
7. Methyl				73
8. Methyl				75
9. Methyl				75
10. Methyl				75
11. Methyl				75
12. Methyl				75

TABLE I (*Continued*)

R'	R"	R'"	−NR"R'"	References
13. Methyl				75
14. Methyl				75
15. Methyl				75
16. Methyl				75
17. *o*-Chlorophenyl			Morpholino	76
18. *o*-Chlorophenyl	Methyl	Methyl		76
19. β-Cyanoethyl	Methyl	Methyl		77
20. β-Cyanoethyl	Isopropyl	Isopropyl		77
21. β-Cyanoethyl			Morpholino	77
22. 2-Methylsulfonylethyl			Morpholino	78
23. *p*-Nitrophenylethyl			Morpholino	79
24. *p*-Nitrophenylethyl				75
25. *p*-Nitrophenylethyl				75
26. *p*-Nitrophenylethyl				75
27. 2,2,2-Trichloroethyl	Methyl	Methyl		80
28. 1,1-Dimethyl-2,2,2-trichloroethyl	Methyl	Methyl		80
29. 1,1-Dimethyl-2,2,2-trichloroethyl	Isopropyl	Isopropyl		80
30. 1,1-Dimethyl-2,2,2-trichloroethyl			Morpholino	80
31. 1,1-Dimethyl-2-cyanoethyl			Morpholino	81

certain 5-substituted tetrazoles such as 5-*p*-nitrophenyltetrazole[83] and 5-trifluoromethyltetrazole[80] as well as hydroxybenzotriazole,[78] *N*-methylimidazole hydrochloride,[80] and *N,N*-dimethylaminopyridine hydrochloride[80] all activate **28e** more effectively than tetrazole. Matteucci and Froehler have concluded from ^{31}P NMR studies and the relationship between tetrazole acidity and activation that protonation of **28e** is rate-determining in the phosphitylation reaction.[83] Recently several additional phosphoramidites having higher membered cyclic amine structures (Table I) have been synthesized, shown to be isolatable by silica gel column chromatography, and are stable for months as dry foams.[75] These compounds have been used successfully for preparative, solution synthesis of dinucleotides where low excesses of phosphoramidites are desirable. Currently, however, **28f** is the preferred reagent primarily because it is not only stable as a solid or in solution (without silica gel column chromatography), but it is also readily activated by tetrazole, a commercially available, nonhygroscopic solid, to form internucleotide phosphite linkages in less than a minute under normal reaction conditions.[84]

Most phosphorus-protecting groups that have been used with the phosphate diester and phosphate triester approaches have now been applied successfully to the phosphite triester approach. Thus, either in solution or on polymer supports, the 2,2,2-trichloroethyl,[11,17] 1,1-dimethyl-2,2,2-trichloroethyl,[15,80] methyl,[24,35,36,71] 2,2,2-tribromoethyl,[16,26] benzyl,[26] *o*-chlorophenyl,[11,76] *p*-chlorophenyl,[26] β-cyanoethyl,[26,77] *p*-nitrophenylethyl,[26,79] 2-methylsulfonylethyl,[78] and 1,1-dimethyl-2-cyanoethyl[81] groups were examined. For polymer-supported synthesis, initial work using nucleoside chloridites[25,36] and phosphoramidites[72] focused on the methyl-protecting group because it was stable throughout DNA synthesis and could be removed rapidly and efficiently by deblocking with ammonium hydroxide, amines, or preferably thiophenol (see, for example, Tanaka and Letsinger[51]). Recently, however, using long reaction times and extremely basic conditions, Gao *et al*. have shown that methyl-protected deoxyoligonucleotides in the triester form (after oxidation of the methoxyphosphite) are agents potentially capable of methylating N-1 on thymine.[85] These results suggest that DNA prepared using the methyl-protecting group and standard reaction conditions[72,86] may contain a small percentage of N^1-methylthymine. This methylated DNA should not interfere with most biochemical or biological applications of synthetic deoxyoligonucleotides (probing gene libraries, sequencing DNA, cloning experiments with synthetic DNA) but may create problems for those using synthetic DNA for biophysical studies. However for these studies, the synthetic DNA should be rigorously purified not only to remove any methylated DNA but also depurinated or incompletely deprotected DNA as well. Therefore, although the methyl group is compatible with all current uses of DNA if certain precautions are followed for biophysical studies, a long-term objective must be to identify a better phosphorus-protecting group.

Until recently, the 2,2,2-trichloroethyl and 1,1-dimethyl-2,2,2-trichloroethyl-protecting groups were unsatisfactory for use with polymer supports because removal was only possible via reductive cleavage in heterogeneous solutions.[25] These are not conditions which can be adapted to solid-phase synthesis where removal of the phosphate-protecting group must take place prior to basic cleavage of the DNA from the support. Otherwise the basic conditions will lead to hydrolysis of internucleotide triester linkages and degradation of the product. Recently, however, certain phosphines [triphenylphosphine, tributylphosphine, and tris(dimethylamino)phosphine] have been shown to remove trihaloester groups bound to a solid support.[16] The possibility therefore exists that 2,2,2-trichloroethyl and 1,1-dimethyl-2,2,2-trichloroethyl groups can be used to protect phosphorus during polymer support synthesis without alkylation of thymine. Unfortunately deblocking conditions are rather rigorous (80°C, 3 hr). Thus, these conditions might lead to side product formation or incomplete deprotection of longer DNA segments. Careful investigations will have to be completed to check these undesirable possibilities.

To simplify procedures for deblocking fully protected deoxyoligonucleotides, Sinha *et al.*[77] and also Schwarz and Pfleiderer[75] showed that the β-cyanoethyl and *p*-nitrophenylethyl groups, which had previously been used in phosphite chemistry,[26] could also function as phosphorus-protecting groups when part of phosphoramidite synthons (Table I). Both of these groups are quite attractive because they are removed rapidly with concentrated ammonium hydroxide, the same reagent used to cleave the deoxyoligonucleotide from the support and to deacylate the base amino groups. Moreover, since these groups are linked to phosphate through a methylene group rather than a methyl, alkylation of thymine should be less of a problem. However, the β-cyanoethyl group has been shown to be unstable to *N,N*-dimethylaminopyridine, an important reagent in the capping solution.[84] This means that high coupling yields can be maintained throughout all synthesis cycles only by increasing the concentration of nucleoside phosphoramidite as the synthesis progresses. Perhaps the most promising base-labile phosphorus-protecting group is the *p*-nitrophenylethyl. Because of its increased stability toward base, it should resist cleavage by *N,N*-dimethylaminopyridine but still be susceptible to relatively mild basic deprotection reagents.

Several routes, as outlined in Fig. 7, have been developed for the synthesis of deoxynucleoside phosphoramidites. The first three pathways[72,73]

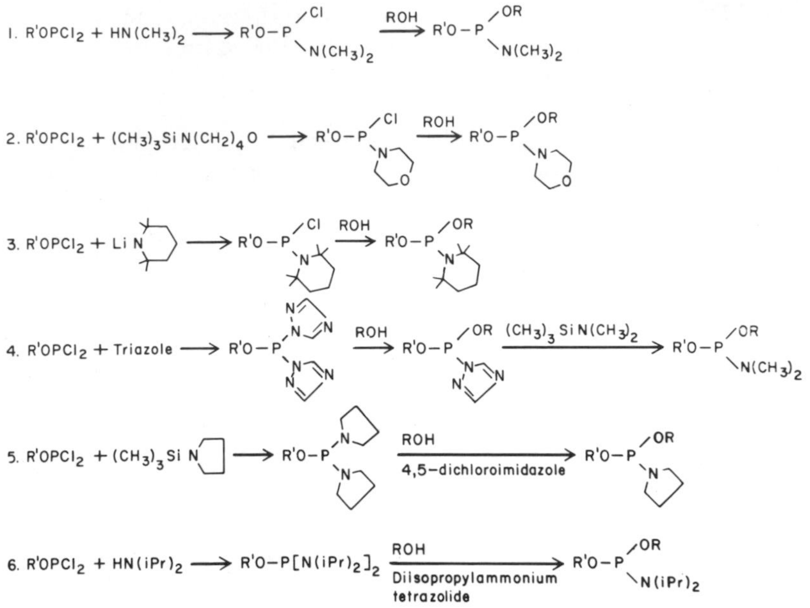

1. $R'OPCl_2 + HN(CH_3)_2 \longrightarrow R'O-P\begin{smallmatrix}Cl \\ N(CH_3)_2\end{smallmatrix} \xrightarrow{ROH} R'O-P\begin{smallmatrix}OR \\ N(CH_3)_2\end{smallmatrix}$

2. $R'OPCl_2 + (CH_3)_3SiN(CH_2)_4O \longrightarrow R'O-P\begin{smallmatrix}Cl\end{smallmatrix} \xrightarrow{ROH} R'O-P\begin{smallmatrix}OR\end{smallmatrix}$

3. $R'OPCl_2 + Li\,N \longrightarrow R'O-P\begin{smallmatrix}Cl\end{smallmatrix} \xrightarrow{ROH} R'O-P\begin{smallmatrix}OR\end{smallmatrix}$

4. $R'OPCl_2 + Triazole \longrightarrow R'O-P \xrightarrow{ROH} R'O-P\begin{smallmatrix}OR\end{smallmatrix} \xrightarrow{(CH_3)_3SiN(CH_3)_2} R'O-P\begin{smallmatrix}OR \\ N(CH_3)_2\end{smallmatrix}$

5. $R'OPCl_2 + (CH_3)_3Si\,N \longrightarrow R'O-P \xrightarrow[\text{4,5-dichloroimidazole}]{ROH} R'O-P\begin{smallmatrix}OR\end{smallmatrix}$

6. $R'OPCl_2 + HN(iPr)_2 \longrightarrow R'O-P[N(iPr)_2]_2 \xrightarrow[\substack{\text{Diisopropylammonium} \\ \text{tetrazolide}}]{ROH} R'O-P\begin{smallmatrix}OR \\ N(iPr)_2\end{smallmatrix}$

Fig. 7. Synthesis pathways for nucleoside phosphoramidites. R' is as defined in Table I. R is an appropriately protected nucleoside or deoxynucleoside; iPr indicates isopropyl.

involve the intermediate synthesis of dialkylaminochlorophosphines which are isolated by fractional distillation and then condensed with appropriately protected nucleosides to form the nucleoside phosphoramidites (**28d–f,** also see Table I. The most convenient aspect of this synthesis is the workup of the product since this can most conveniently be accomplished simply by an aqueous extraction followed by precipitation of the product as a colorless, stable solid. In some cases, such as with the morpholidite or several cyclic alkylamino phosphoramidites, the products can also be further purified by conventional silica gel column chromatography. Even when isolated without column chromatography, **28f** is relatively free of side products and stable to hydrolysis and oxidation.[73] However, this route suffers from certain undesirable features. For example, since the intermediate chlorodialkylaminophosphines are extremely pyrophoric and easily oxidized or hydrolyzed, they must be stored in tightly sealed containers under an inert gas atmosphere. Additionally, the final product can be unstable if not isolated free of the amine hydrochloride which is generated during formation of the nucleoside phosphoramidite. This can be done most conveniently by a careful extraction procedure.[72] The amine hydrochloride can also cause variable amounts of hydrolysis to the phosphinic acid during the aqueous extraction. This is because the different phosphoramidites have variable stability toward hydrolysis with amine hydrochlorides (or activation).[73] Additionally contact time between amine hydrochlorides and phosphoramidites will vary because the different nucleoside phosphoramidites will partition into the organic phase at rates that will be a function of the hydrophobicity of the *N,N*-dialkylamino group. A final feature of these reaction schemes is the presence of a small (relative to the amount produced from using the dichlorophosphine) but undesirable yield of $3' \rightarrow 3'$ dinucleotide. This product is generated by further reaction of amine hydrochloride with the phosphoramidite which then reacts with nucleoside present in the reaction mixture.

Because of these limitations, bisaminophosphines were examined as potential reagents for synthesizing nucleoside phosphoramidites. Initially very reactive compounds such as methylphosphoroditetrazolide[70,71] or the corresponding bistriazoylphosphine[70,76] were investigated. Using the latter reagent, nucleoside phosphoramidites were synthesized via a one-pot procedure (Scheme 4, Fig. 7) and isolated by aqueous extraction and precipitation. Compound **28e**, in yield exceeding 90%, was free of the $3' \rightarrow 3'$ dimer. However formation of **28d** and also **28e** or **28d** where methyl is replaced with *o*-chlorophenyl leads to significant generation of the $3' \rightarrow 3'$ dimer (15%). Schemes 5 and 6 offer an alternative method.[87,88] The first step is to synthesize the bis-*N,N*-dialkylaminophosphine, which can be isolated by fractional distillation. In contrast to the other phosphitylating agents that have been proposed (dichlorophosphites, chlorodialkylaminophosphites, bistetrazolyl phosphites, and bistriazoylyl phosphites), these intermediates are stable to normal laboratory conditions such as oxidation or hydrolysis and can be simply stored in a closed container without special precautions. The next step is to activate the bis-*N,N*-dialkylaminophosphine with acids such as 4,5-dichloroimidazole[87] or the tetrazolide salt of the *N,N*-dialkylamine that corresponds to the bis-*N,N*-dialkylaminophosphine.[88] In Scheme 6, this salt would be diisopropylammonium tetrazolide (morpholinium tetrazolide has been used to activate bismorpholinomethoxyphosphine[88]). With 4,5-dichloroimidazole, the phosphoramidite is obtained in high yield (86%) but is contaminated with approximately 9% of the $3' \rightarrow 3'$ dimer. The tetrazolide salt, however, appears to be a superior activating agent because the product which is obtained in high yield is free of the two most common side products, the deoxydinucleoside $3' \rightarrow 3'$-triphosphite and the deoxynucleoside $3'$-phosphinic acid. Apparently diisopropylammonium tetrazolide can only protonate (activate) bis-*N,N*-diisopropylaminophosphine and consequently form the deoxynucleoside $3'$-phosphoramidite. Unlike acids such as amine hydrochlorides and 4,5-dichloroimidazole, diisopropylammonium tetrazolide is not sufficiently acidic to further activate the deoxynucleoside phosphoramidite to form either the 3′ → 3′ dimer from excess deoxynucleoside or the deoxynucleoside 3′-phosphinic acid from trace water contamination in solvents. Synthesis via activation with tetrazolide salts also appears to be superior because acidic amine hydrochlorides are not produced during synthesis, as is the case with chlorophosphites. Consequently acidic amine hydrochlorides cannot catalyze hydrolysis of the deoxynucleoside phosphoramidites during the aqueous workup or coprecipitate with the product which would lead to its decomposition upon storage. Because of these considerations, the deoxynucleoside phosphoramidites derived by synthesis from bis-N,N-dialkylaminophosphines activated with the corresponding tetrazolide salts and isolated via a simple aqueous extraction are very stable and essentially homogeneous without column chromatography.[88] Based on these various results, Scheme 6 is now the preferred method for synthesizing nucleoside phosphoramidites.

Because bis-N,N-dialkylaminomethoxyphosphines were shown to be excellent phosphitylating agents, a new method for preparing DNA was introduced independently by Beaucage[87] and also Barone, Tang, and Caruthers.[88] The key feature of this method was the synthesis *in situ* of deoxynucleoside phosphoramidites. The reaction sequence begins by reacting a protected deoxynucleoside with bisdiisopropylaminomethoxyphosphine using diisopropylammonium tetrazolide as activating agent (Fig. 8). This salt is preferred because activation is rapid and leads to high yields of the deoxynucleoside phosphoramidites without production of side products. Bismorpholinomethoxyphosphine and bispyrrolidinomethoxyphosphine in combination with morpholinium tetrazolide and

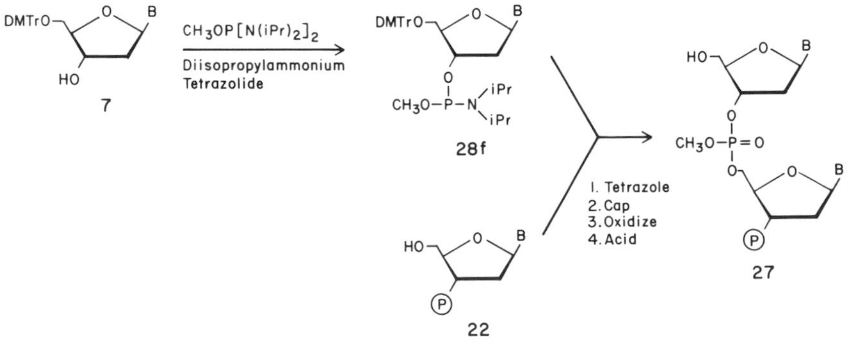

Fig. 8. *In situ* DNA synthesis.

4,5-dichloroimidazole, respectively, have also been examined.[87,88] The deoxynucleoside phosphoramidites thus formed (**28f**) are used directly without isolation for synthesizing deoxyoligonucleotides on silica-based supports. Thus, all four deoxynucleoside phosphoramidites are prepared *in situ* prior to DNA synthesis, activated by addition of tetrazole, and added in the appropriate sequence to the support to form a deoxyoligonucleotide. It is important to consume all the bis-*N,N*-diisopropylamino-methoxyphosphine during preparation of **28f**. This can most easily be done by using excess protected deoxynucleoside[88] or quenching the reaction with water.[69] Otherwise the excess bis-*N,N*-diisopropylamino-methoxyphosphine will react with tetrazole during the condensation step and, in this activated form, phosphitylate the growing DNA segment at the available 5'-hydroxyl group. This phosphitylation effectively caps the growing deoxyoligonucleotide on the support and reduces the overall yield. The approach, as outlined in Fig. 8, is especially useful to the nonchemist because the pathway eliminates many of the pitfalls common to most DNA synthesis procedures. For example the synthons, protected deoxynucleosides and silica-linked deoxynucleosides, are nonhygroscopic solids that can be stored indefinitely. Additionally, the phosphitylating reagent, bis-diisopropylaminomethoxyphosphine, is stable under normal laboratory conditions to oxidation or hydrolysis and the diisopropylammonium tetrazolide is a nonhygroscopic, easily prepared reagent. Thus, the starting materials require no special handling or storage precautions. Also of importance to the nonchemist is that the bis-*N,N*-diisopropylaminomethoxyphosphine reacts rapidly with extraneous water when activated with diisopropylammonium tetrazolide to form phosphoamidous acid, a compound which is inert to DNA synthesis reagents and therefore not deleterious to the chemistry. Thus, if the researcher uses slightly wet solvents or fails to remove trace water from reagents such as protected deoxynucleosides, bisdiisopropylaminophosphine serves as its own desiccant by reacting with this extraneous water.

E. Assembly of Deoxyoligonucleotides
on Solid-Phase Supports

The synthesis of DNA on solid supports using the phosphite approach proceeds with appropriately protected deoxynucleoside 3'-phosphoramidites as synthons. The process generally follows the procedures outlined schematically in Fig. 6 when previously isolated synthons are used or in Fig. 8 when deoxynucleoside phosphoramidites are prepared *in situ*. Irrespective of which variation is followed, the same series of chemical steps as outline in Table II is used to prepare DNA. Thus, each synthesis cycle

TABLE II
Chemical Steps for One Synthesis Cycle

Step	Reagent or solvent[a]	Purpose	Time (min)[b]
1	Dichloroacetic acid in CH_2Cl_2 (2 : 100, v/v)	Detritylation	3
2	CH_2Cl_2	Wash	0.5
3	Acetonitrile	Wash	1.5
4	Dry acetonitrile	Wash	1.5
5	Activated nucleotide in acetonitrile[c]	Add one nucleotide	5
6	Acetonitrile	Wash	0.5
7	DMAP/THF/lutidine[d] (6 : 90 : 10, w/v/v) 0.1 ml acetic anhydride	Cap	2
8	THF/lutidine/H_2O (2 : 2 : 1, v/v/v)	Wash	1
9	I_2 Solution[e]	Oxidation	1
10	Acetonitrile	Wash	0.5
11	CH_2Cl_2	Wash	0.5

[a] Multiple washes with the same solvent involve filtration between wash steps. Each step volume is 1 ml unless indicated.

[b] The total time per cycle for manual synthesis is usually about 15–17 min as described. However most machines, through optimization of various parameters, have a cycle time of 7–10 min using the same reagents.

[c] For each μmole of deoxynucleoside attached covalently to silica gel, 0.4 M tetrazole (0.2 ml) and 0.1 M deoxynucleoside phosphoramidite (0.2 ml) are premixed in acetonitrile.

[d] DMAP, Dimethylaminopyridine; THF, tetrahydrofuran.

[e] THF/lutidine/H_2O (2 : 2 : 1, v/v/v) containing 0.2 M iodine.

begins by treatment of **21** first with dichloroacetic acid to remove the trityl ether followed by washes with dichloromethane, the solvent used during the detritylation step, and acetonitrile in preparation for the condensation step. The next step proceeds by first converting the protected deoxynucleoside 3'-N,N-diisopropylaminophosphoramidite (**28f**) to the tetrazolide and then adding this activated nucleotide to the support. This reaction should be carried out using excess amidite to insure complete and rapid reaction (less than 1 min) and to account for some amidite hydrolysis due to unavoidable trace water contamination through solvents and glassware. After washing with acetonitrile to remove excess synthon, the support containing the phosphite is sequentially treated with capping reagents, an aqueous hydrolysis solution, and finally an aqueous iodine solution to oxidize it to the phosphate. Thus, the synthesis proceeds stepwise in a 3' → 5' direction by the addition of one nucleotide per cycle. Because the current coupling efficiencies are so very high (greater than 99.5%), there is no need to use dinucleotide or trinucleotide synthons for preparing even long deoxyoligonucleotides (50–100 mononucleotides each).[89] There have, however, been reports where dinucleotide synthons carrying a 3′-activated phosphite were used for solution[16] and solid-phase support synthesis.[90] However because of the additional effort required to prepare and purify dimer synthons, they will probably not be useful for routine DNA synthesis. Their utility will undoubtedly be for synthesis where a deoxyoligonucleotide analog is needed. For example, deoxydinucleotides containing an intermediate methyl phosphonate linkage were prepared in solution, the stereoisomers resolved by silica gel column chromatography, and then used as dimer units containing a 3-phosphoramidite to prepare chiral *lac* operator segments for studies on the *lac* operator–*lac* repressor interaction.[91]

The cycle, as outlined from the earliest publications using deoxynucleoside phosphoramidites, has recommended an aqueous wash and preferably even the acetic anhydride capping step prior to iodine oxidation.[43,45,72,73,86,92] This was because the protected bases were observed to react transiently with activated phosphites to form phosphite esters and amidites. These adducts were then easily hydrolyzed from the bases without modifying the nucleotides simply via a hydrolytic wash.[69,72,73,87,88] Of additional significance was the expectance that oxidation of phosphite adducts to phosphate would stabilize the resulting phosphate triesters and lead to base modifications during deprotection. Over the years, ample evidence has accumulated to demonstrate that numerous side products accumulate when stable adducts form between nucleotide bases and either aryl sulfonyl condensing agents[93–96] or phosphorylating reagents.[97–100] Recently attempts have been made to shorten this reaction cycle by carrying out the iodine oxidation immediately after the condensation step and deleting the aqueous hydrolytic wash.[101,102] Unfortunately these alterations in the cycle lead to modification of the guanine base and a requirement to protect the O-6 of guanine during synthesis. This protection step complicates the synthesis and also requires an additional deblocking step. Incorporation of the aqueous wash and the capping steps prior to oxidation thus appears to be the most reasonable solution to a problem that does not exist unless oxidation immediately follows the condensation step.

Several manual devices are available for the synthesis of DNA using the chemistry outlined in Table II. The simplest utilize test tubes[43,64] or sintered glass funnels.[43,45] If test tubes are used as reaction vessels, each synthesis begins by placing the solid support in the test tube (compound **21**). Reagents and solvents are then added sequentially and removed after a low-speed centrifugation by decantation of the liquid phase. Sintered glass funnels containing **21** are attached to a water aspirator. Reagents and solvents are then added sequentially and removed by filtration. Although these manual methods are extremely simple and convenient for the nonchemist, certain precautions are necessary. One is to use an inert gas atmosphere over the reaction vessel during the condensation step to reduce hydrolysis and oxidation of the activated phosphite. This is most conveniently done by placing a rubber septum attached to an inert gas source over the reaction vessel. Another precaution is to use a larger excess of the activated deoxynucleoside phosphite to insure that hydrolysis by trace water contamination does not quench the synthon. This is necessary since manual devices, which are usually open to the atmosphere, are more easily contaminated with water. Manual synthesis usually suffers from the possibility of human error as well. Thus, to insure that the incorrect deoxynucleoside phosphoramidite is not added to the growing DNA segment, a series of color-coded synthons were developed.[103] As can be seen by examining the reaction scheme shown in Fig. 6, a dimethoxytrityl group is removed during each synthesis cycle. More importantly, this trityl group is part of the mononucleotide synthon added to the support during each synthesis cycle. Since this trityl group forms an orange color during release from the support by acid, it is diagnostic of the synthon added at each step. This principle was used to develop four trityl groups which generated different colors in acid with one being assigned to each of the nucleotides (28e). Thus, by observing the color produced during each acid detritylation step, the sequence of the DNA segment being synthesized could be checked in a systematic manner.

Another extremely simple device for use in manual synthesis is the syringe.[51] Silica or CPG bearing a covalently linked deoxynucleoside is placed in a syringe equipped with a filter at the base. Synthesis is carried out by successively drawing solutions into the syringe and then expelling them. Synthesis using a syringe usually requires a smaller excess of the activated deoxynucleoside phosphite since the silica support is sealed within the reaction chamber and therefore less accessible to contamination with water. However the operator usually is restricted with a syringe to one synthesis at a time rather than the simultaneous preparation of six to eight deoxyoligonucleotides which is possible using test tubes and sintered glass funnels.

Also available are several semiautomated and completely automated machines for synthesizing DNA. Numerous designs for semiautomated machines have been proposed.[25,35,36,43,46,104,105] The solid support is placed in a column joined through a series of valves and tubing to a pump and an injector loop. Usually reagents and solvents are attached to the machine and activated deoxynucleotide synthons are added through the injector loop. The machine shuts down after one cycle, and the operator then recycles the program.

Recently Frank *et al.* proposed a semiautomated device using four columns to synthesize simultaneously a large number of DNA segments.[106] The approach was initially developed for use with phosphate triesters but has now also been shown compatible with the phosphite triester approach.[107] Generally a deoxynucleoside attached to a cellulose disk is placed in a column and extended by one nucleotide using the cycle shown in Table II. By designating separate columns for addition of deoxyadenosine, thymidine, deoxycytidine, and deoxyguanosine phosphoramidites and using several disks concurrently, mononucleotides can be added to a large number of segments. The disks are then resorted according to the sequences of the segments being synthesized and the cycle repeated. In contrast to packing a particulate matrix in a column and synthesizing one DNA segment, this approach uses four columns to synthesize many segments simultaneously. The approach is particularly attractive for completing mutagenesis studies, including base pair changes and deletions. The operator simply completes multiple syntheses of the same segment but splits the disks at the mutagenesis site for the appropriate changes in sequence. (Adding all four or some combination of mononucleotides simultaneously to the growing segment is not advisable because the purity and therefore the reactivity of each mononucleotide can vary with different preparations, irrespective of which chemistry, phosphite or phosphate triester, is used.) With manual or semiautomated approaches involving silica, the same mutagenesis is most conveniently completed by carrying out one synthesis up to the mutation site, splitting the silica and completing multiple syntheses at the locus of mutations, and then recombining the silica for the remaining synthesis cycles. The choice between these two approaches depends on the subsequent cloning procedures. The filter disk approach provides resolved, mutated segments with the least number of manipulations whereas the latter method yields the same segments using minimal synthesis steps but as a mixture having equal portions of all mutated oligomers. Thus, if no biological selection is available, the former approach is preferred. Conversely, the latter should be used when a selection procedure can be followed.

Completely automatic DNA synthesizers have also been described.[50,108] Automatic gene machines that are available commercially can usually be programmed to synthesize deoxyoligonucleotides containing at least 100 mononucleotides[89] or segments containing more than one deoxymononucleotide at a predetermined position (as mutagenesis primers or as probes for gene libraries). For the synthesis of mixed sequence probes or primers however, these machines add mixtures of mononucleotides at heterologous positions. If having equivalent mixtures of all possible heterologous segments is important, the operator must therefore rely upon the manufacturer for equal mixing of mononucleotides in the machine and for having all synthons of the same purity.

F. Deblocking, Purification, and Characterization of Synthetic Deoxyoligonucleotides

1. Deblocking

Essentially all functional groups of the nucleotides are protected to avoid side reactions and to increase the solubility of the synthons. Thus, the deoxyoligonucleotides as synthesized on the solid silica support contain numerous protecting groups that must be removed completely before any biochemical or biological studies are possible. Additionally these protecting groups should be removed in a specific sequence to minimize the production of side products. For example, the first step is conversion of the phosphate triester to the phosphate diester because the diester is much less susceptible to the reagents used to remove the nucleotide exocyclic amino-protecting groups. This step is followed by removal of the base-protecting groups and then treatment with acid to remove the 5'-dimethoxytrityl ether. This is because it is well known that deoxyadenosine is more resistant to depurination in acid than is 6-amino-protected deoxyadenosine. One could conceivably remove the dimethoxytrityl ether before hydrolysis from the support but, in cases where maximum purification is desired, this choice eliminates the possibility of using reverse-phase chromatography. Additionally, a free 5'-hydroxyl, which would be generated after detritylation of a support-bound triester, has in some cases been shown to attack internucleotide phosphate triesters and lead to rearranged deoxyoligonucleotides.[109] Thus, the preferred order of deprotection is first the phosphate-, then base-, and finally dimethoxytrityl-protecting groups.

The reagents used for deblocking will depend upon the protecting groups. If phosphorus contains a methyl group, triethylammonium thiophenoxide is the reagent of choice to convert the triester to the diester. The next step is to treat with concentrated ammonium hydroxide to cleave the deoxyoligonucleotide from the support and to remove exocyclic amino-protecting groups.[25] This is an extremely critical step. If the ammonium hydroxide is improperly stored for some time at room temperature in a loosely sealed bottle, free ammonia will escape and the resulting dilute ammonia solution will not remove all protecting groups. Concentrated ammonium hydroxide should be stored tightly sealed in a freezer and opened only briefly during transfers to DNA samples. An alternative deprotection scheme involves initial treatment with concentrated ammonium hydroxide to remove the methyl- or β-cyanoethylphosphorus-protecting group, the base-protecting groups, and the deoxyoligonucleotide from the support. These are currently the two most popular deblocking schemes, but they will undoubtedly change as the various protecting groups are replaced. For example, a series of exocyclic amino-protecting groups that could be deblocked by a nonvolatile but easily removable reagent would be desirable. In this way, incomplete deprotection due to the volatility of aqueous ammonia would cease to be a source of side products.

2. Purification

The final product from the synthesis and deblocking of a DNA segment as prepared on a solid phase contains the desired deoxyoligonucleotide and a series of side products including truncated failure sequences, depurinated DNA, incompletely deprotected DNA, and minor uncharacterized impurities. Ideally these side products should be removed during purification. Although this is indeed the case for short segments (10–20, mononucleotides), the current solid-phase methodologies can yield segments containing 50–150 monomers where purification to homogeneity is beyond the capabilities of current methods. For most biochemical experiments, trace impurities are not important. These applications include using synthetic DNA as probes for gene libraries, as primers for DNA sequencing, and for cloning experiments. In the latter case, impurities leading to mutations are eliminated by selecting a correct segment through DNA sequencing of the cloned DNA. However, for certain biophysical or biochemical experiments, rigorous purification is necessary and all the available techniques should be used so that maximum purity can be obtained.

The simplest and one of the most effective purification processes is polyacrylamide gel electrophoresis. This technique as originally adapted to purification of synthetic DNA[43,45] involves loading the reaction mixture onto a polyacrylamide gel equilibrated with 7 M urea, resolving the product from shorter sequences containing as little as one charge difference, and visualizing the product with ultraviolet light. The acrylamide containing the product is then cut from the gel and the deoxyoligonucleotide is eluted using standard procedures.[110] Two main limitations are low loading capacities and poor resolution as the length of the segment increases (usually 40–50 mononucleotide length). The latter problem is of importance to those who use synthetic DNA for cloning experiments and DNA sequencing. If the product eluted from the gel contains unresolved failure sequences or partially deprotected product, then families of deletion and base pair altered mutations will be observed upon cloning. The main advantage of this approach is speed since many deosyoligonucleotides can be purified simultaneously and most biologists, biochemists, or molecular biologists are familiar with this procedure.

High-performance liquid chromatography is the major alternative method for purifying synthetic DNA. When ion-exchange column chromatography is used, the desired product is always eluted later than any shorter sequences and can therefore be easily identified. Generally an anion-exchange column such as Partisil 10 SAX (Whatman) is loaded with a reaction mixture and the deoxyoligonucleotides fractionated using salt gradients in formamide. The product peak can then be desalted and used for biochemical studies. If extremely pure DNA is required, the isolated product can be further purified on a reverse-phase column.[111,112] Alternatively, as can be seen by examining the reaction scheme outlined in Fig. 6, the only deoxyoligonucleotide containing the 5'-dimethoxytrityl group is the product. The shorter truncated segments would not have this group if both the detritylation and capping for each cycle go to completion. Therefore if all protecting groups except the 5'-dimethoxytrityl are removed prior to purification, then the product should be retained exclusively on a reverse-phase column and eluted from truncated failure sequences using acetonitrile gradients. This was indeed observed and can be used as a very effective purification procedure.[25] However if the product is incompletely freed of protecting groups, depurinated, or alkylated, these side products will not be fractionated from undamaged product unless further purified on either a reverse-phase or an anion-exchange column after removal of the trityl ether. For all these column purification procedures, a serious limitation is loss of resolution in the 25- to 30-nucleotide range. A major advantage over polyacrylamide gel electrophoresis is that milligram quantities can be purified by columns.

3. Characterization

The final step of synthesis is characterization of the product by sequencing methods. The two most useful are the wandering spot procedure[113,114] and the Maxam and Gilbert sequencing method[110] as adapted to sequencing small synthetic segments.[115] The former procedure is most useful for analyzing purity, as modified or partially protected deoxyoligonucleotides are easily resolved from the homogeneous product and thus give an estimate of purity. However, the procedure is extremely difficult to execute. The latter method is rapid and easily gives the base sequence. It lacks sensitivity for judging purity.

IV. SOLID-PHASE PHOSPHITE TRIESTER SYNTHESIS OF RNA

Methodologies for the chemical synthesis of oligoribonucleotides on solid-phase supports have not kept pace with advances in DNA synthesis.[14] Clearly the reason is the problem of selectively protecting and deprotecting an additional nucleophilic center, the 2'-hydroxyl group. Only recently, as summarized in Fig. 9 and 10, have methods been proposed that lead to the synthesis on polymer supports of oligoribonucleotides containing 10–20 mononucleotides. These procedures differ primarily in the strategies used to solve problems associated with the chemical requirements of a 2'-hydroxyl-protecting group.

One approach,[116-123] which follows from earlier research on DNA synthesis,[25,36,45,72] is outlined in Fig. 9. Thus, an appropriately protected ribonucleoside was covalently linked to silica through either the 2'- or 3'-hydroxyl group (29a–c). After removal of the trityl ether (dimethoxytrityl or monomethoxytrityl) to form 30a–c, compound 33a–f was condensed to

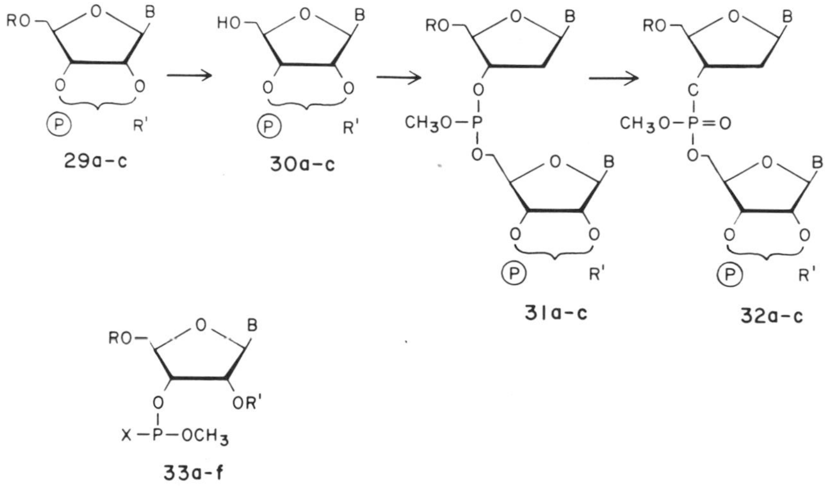

Fig. 9. Synthesis of RNA 3' → 5' on polymer supports. Abbreviations: **29a–c**, R = MMTr and R' = TBDMS **(a)**, R = DMTr and R' = Bz **(b)**, R = DMTr and R' = THP (c); **33a–f**, R = MMTr and R' = TBDMS and X = Cl **(a)**, R = DMTr and R' = Bz and X = DMA **(b)**, R = DMTr and R' = THP and X = DMA **(c)**, R = DMTr, R' = THP, and X = DIPA **(d)**, R = MMTr, R' = TBDMS, and X = DIPA **(e)**, and R = MMTr, R' = TBDMS, and X = morpholino **(f)**. DMTr, Dimethoxytrityl; MMTr, monomethoxytrityl; TBDMS, *tert*-butyldimethylsilyl; Bz, benzoyl; THP, tetrahydropyranyl; DMA, *N,N*-dimethylamino; DIPA, *N,N*-diisopropylamino; B is uracil or appropriately protected cytosine, adenine, and guanine.

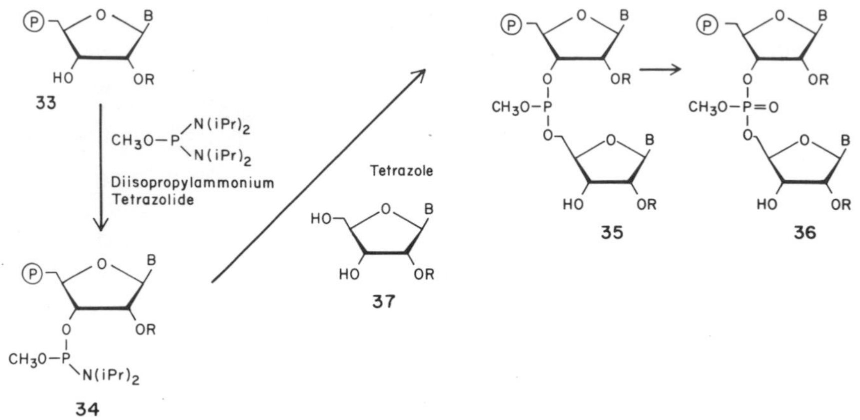

Fig. 10. Synthesis of RNA 5′ → 3′ on polymer supports. Abbreviations: R = tetrahydropyranyl; B and iPr are as defined in the legend to Fig. 9.

yield **31a–c**. After capping and oxidation to generate the dinucleotide (**32a–c**), the synthesis steps can be recycled and an oligoribonucleotide generated. Initially, appropriately protected ribonucleoside 3′-chlorophosphites (**33a**) were used to synthesize compounds as large as a heptadecadeoxyoligonucleotide.[116-119] As was found earlier in the DNA series,[25,36,72] these phosphites are moisture sensitive, difficult to handle, and generally lead to lower yields. Research has therefore focused on using the N,N-dimethylamino-,[120,121] N,N-diisopropylamino-[122,123] and morpholinophosphoramidites[122] (**33b–f**) as synthons. The results have been especially encouraging with **33e** where the synthon contains a 2′-silyl-protecting group.[122] The yields per condensation (95–98%) approached those observed in DNA synthesis. The problem however is the 2′-protecting group. The silyl is somewhat controversial as a 2′-protecting group[124-126] since in some cases silyl migration 2′ to 3′ has been observed during either purification of the protected nucleoside or removal of this protecting group. A related problem is incomplete chromatographic separation of 2′-and 3′-tert-butyldimethylsilyl-protected guanosine. Recently however it has been reported that when triisopropylsilyl is substituted for tert-butyldimethylsilyl, the 2′- and 3′-silylated guanosines are most easily resolved.[117] Other 2′-protecting groups have also been used, such as the benzoyl[120] and tetrahydropyranyl groups.[121,123] Benzoyl leads to 1–4% isomerization per cycle which is unsatisfactory for a repetitive synthesis. The tetrahydropyranyl group is partially hydrolyzed during the cyclical detritylation step which leads to internucleotide bond cleavage. As a consequence the low yields obtained with tetrahydropyranyl limits the usefulness of this group to the synthesis of octa- to dodecanucleotides. These various observations point to potential problems with each of the three 2'-protecting groups so far used with activated phosphites and suggest that additional research is necessary to develop either a more satisfactory 2'-protecting group or a different synthesis strategy.

Among the nucleoside phosphoramidites tested, the N,N-diisopropyl derivative (33e) yields the best results. Of interest was that the morpholidite (33f) was more difficult to form and also reacted much more slowly to form 31a when compared to the deoxynucleoside derivative.

A different strategy for synthesizing RNA is outlined in Fig. 10. The approach involves initially condensation of 33, a nucleoside attached covalently to silica or controlled pore glass through an ester bond, with excess bis-diisopropylaminomethoxyphosphine using diisopropylammonium tetrazolide as catalyst[123] to form a nucleoside 3'-phosphoramidite on the support (34). Since N,N-diisopropylaminophosphoramidites are stable to water, there is no problem of hydrolysis under these reaction conditions. Synthesis of compound 34 in high yield requires 1 hr, a rate which is approximately the same as for the deoxy series.[88] Once the formation of 34 is complete, the solution is filtered from the support and a second solution containing excess 2'-protected nucleoside (37) and tetrazole is added. Tetrazole, being much more acidic than the tetrazolide salt, activates the phosphoramidite on the support which leads to regioselective condensation with the 2'-protected nucleoside to form a $3' \rightarrow 5'$ dinucleoside phosphite (35) attached to the support. By adding part of an equivalent of bisdiisopropylaminomethoxyphosphine to 37, trace water contamination, which would quench the tetrazolide activation of 34, is removed by conversion to phosphoamidous acid and as a consequence high yields are maintained. Oxidation of phosphite to phosphate, which yields 36, completes the steps required for addition of one nucleotide. Using this procedure, a dodecanucleotide was synthesized in high yield and shown by enzymatic analysis to contain only $3' \rightarrow 5'$ internucleotide linkages. Since no repetitive deprotection steps are incorporated into the strategy, the usual problems associated with identifying compatible protecting groups on the 2'-, 3'-, and 5'-hydroxyls cease to exist. An additional advantage is that 37 is easy to prepare via a one-flask procedure.[127] This approach therefore holds considerable promise for synthesizing RNA on solid supports.

V. SYNTHESIS OF DNA AND RNA ANALOGS USING PHOSPHITE INTERMEDIATES

The phosphite method provides an attractive route for synthesizing oligonucleotides having phosphorus, sugar, or base modifications. One reason for this is that the phosphorus III internucleotide linkage is chemically very versatile and can be converted into several backbone-modified polynucleotides. Additionally nucleoside analogs, which are used to synthesize base-modified oligonucleotides, are available in extremely limited amounts and must be conserved during the synthesis of DNA or RNA segments. These restrictions are most easily met by preparing oligomers on polymer supports using phosphite chemistry where the small amounts needed for biological studies (microgram amounts) can be synthesized in high yield with minimum manipulation of intermediates. For example, the *in situ* approach eliminates the need to isolate nucleoside phosphoramidite synthons. This section discusses the major classes of polynucleotide analogs that have been prepared to date.

The phosphite triester approach as outlined in Fig. 6 and 8 yield internucleotide phosphites (**38a,b**) which are normally oxidized to phosphates (**39a**) (Fig. 11). By selecting various other oxidants, oligonucleotide analogs containing sulfur, nitrogen, selenium, oxygen-17, oxygen-18, and alkylphosphonates can be prepared and deprotected to yield **40a–e.** Thus by using sulfur[128–133] or selenium,[129] mixtures of diastereomers (**39c**) form rapidly and in high yield. Two methods have been used to prepare the Rp and Sp diastereomers of the phosphorothioate-containing oligonucleotides. One involves the preparation of dinucleoside phosphorothioates as diastereomeric mixtures by sulfur oxidation of dinucleoside phosphites, separating the diastereomers by column chromatography, and then adding the chirally pure phosphorothioate dimer to the growing oligonucleotide chain attached to a polymer support.[130,131] Alter-

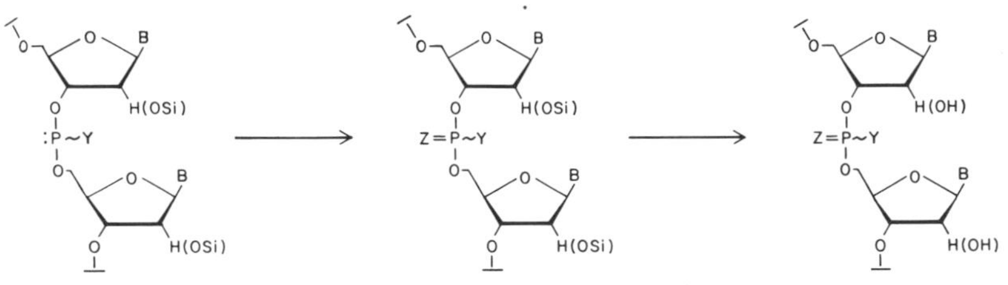

38a, Y = OCH$_3$	39a, Y = OCH$_3$, TCE, Z = O	40a, Y$^-$ = Z = O
b, Y = OTCE	b, Y = OCH$_3$, Z = ^{17}O or ^{18}O	b, Y$^-$ = O, Z = ^{17}O or ^{18}O
c, Y = NEt$_2$	c, Y = OCH$_3$, TCE, Z = S or Se	c, Y$^-$ = O, Z = S or Se
	d, Y = CH$_3$, Z = O	d, Y = CH$_3$, Z = O
	e, Y = OTCE, Z = NH or NEt or NiPr	e, Y$^-$ = O$^-$, Z = NH or NEt or NiPr
	f, Y = OTCE or iPr, Z = O	

Fig. 11. Synthesis of oligonucleotide analogs. Abbreviations: Si, *tert*-butyldimethylsilyl; Et, ethyl; iPr, isopropyl; TCE, trichloroethyl.

natively by a simple manipulation of oxidizing solutions (substitution of sulfur–pyridine for iodine–H_2O), polynucleotides having phosphorothioate substitutions at specific sites can be prepared directly during the stepwise addition of phosphoramidite synthons as outlined in Table II. These diastereomers in some cases can be separated by means of reverse-phase HPLC. Successful separation is dependent upon chain length, base composition, and position of the phosphorothioate linkage within the chain.[132] Recent observations indicate that diastereomers having a 5'-terminal phosphorothioate internucleotide bond and also a 5'-dimethoxy-trityl group can be separated by HPLC.[133]

In the past, phosphorothioate analogs of nucleotides, dinucleotides, and nucleotide polyphosphates have proven to be especially useful for many studies on certain phosphorylytic enzymes.[134] By following the stereochemical course of an enzymatic reaction with a P-chiral substrate, knowledge of whether such a reaction proceeds with retention or inversion of configuration at phosphorus can be obtained. This knowledge in turn provides evidence for or against the existence of a covalent intermediate and therefore limits the number of mechanisms that can be proposed for an enzymatic reaction. Similar studies using phosphorothioate-containing oligonucleotides to examine the course of the *Eco*RI-catalyzed reaction have now been initiated as well. Both the Rp and Sp diastereomers of d[GpGp(S)ApApTpTpCpC], which contain the recognition sequence for the restriction endonuclease *Eco*RI and a phosphorothioate group at the cleavage site, were synthesized, but only the Rp diastereomer was cleaved by *Eco*RI endonuclease. The stereochemical course of this reaction is still under investigation.[130] Phosphorothioate octanucleotides (Rp)- and (Sp)-d[G(p(S)CpG)$_3$p(S)C] and Rp- and Sp-d[C(p(S)GpC)$_3$p(S)G] were also synthesized and used to examine the B → Z transition via circular dichroism spectroscopy.[131] Both (Rp)-d[G(p(S)CpG)$_3$(S)C] and (Sp)-d[C(p(S)GpC)$_3$(S)G] were capable of forming z-type structures at high sodium chloride concentrations. In the case of (Rp)-d[G(p(S)CpG)$_3$p(S)C] where a phosphorothioate of the Rp configuration occurs 5' to a deoxycytidine residue, the B → Z transition is potentiated in comparison to the unmodified oligomer. The remaining two diastereomers retain the B conformation, even at high salt concentration. Further experiments are necessary to understand the basis of these observations.

Oxygen chiral phosphate has been used successfully to assign the resonances in the ^{31}P NMR spectra of several deoxyoligonucleotides.[135–140] This is possible because the quadrupolar oxygen-17 nucleus causes the ^{31}P NMR signal of the phosphate group to which it is bonded to broaden to such an extent that it disappears from the spectrum. Oxygen-18 isotopic substitution is also useful because this isotope causes a small upfield shift on the phosphorus-31 signal. The current most successful method for utilizing these isotope effects in assigning ^{31}P NMR resonances involves regiospecific labeling of the internucleotide phosphate group (P=O) with oxygen-17. This can most easily be done through oxidation of **38a** to **39b** using iodine and oxygen-17-enriched water. Thus, a series of oligonucleotides can be prepared **(40a,b)** which have the same sequence but differ by the position of oxygen-17 enrichment in phosphate. By observing changes in the ^{31}P NMR spectrum of each oligomer due to the reduction of a signal for phosphorus bonded to oxygen-17 ($H_2{}^{17}O$ as available commercially is only 50% enriched in oxygen-17), the unambiguous assignment of each phosphorus signal can be made. An interesting possible route for simultaneously labeling several phosphates using iodine oxidation has also been proposed.[136] By using oxygen-17 and oxygen-18, all three signals from a double labeled tetramer were identified. Therefore by using different ratios of the three oxygen isotopes at each cycle of the phosphate oxidation, it should be possible to assign at least six or more phosphate signals in each oligomer preparation and therefore reduce significantly the amount of synthesis needed to assign completely the phosphorus-31 signals for any oligonucleotide. Iodine oxidation with [^{17}O, ^{18}O]H_2O has also been used in solution chemistry to yield the diastereomeric phosphate triesters which can be resolved by column chromatography.[141,142] These oxygen chiral isotopomers of ^{17}O,^{18}O-labeled dinucleotides can then be incorporated as dimers into oligonucleotides to yield chiral compounds for various biochemical studies.

^{31}P NMR using oxygen-17-enriched phosphate has been used to study drug-DNA interactions and the conformation variations of phosphorus for short DNA duplexes. Thus, complexes of actinomycin D-d(Ap[^{17}O]Gp[^{18}O]Cp[^{16}O]T)$_2$ have confirmed that this drug intercalates between the GpC stacked based pairs.[137] Other research with d(CGCGAAT-TCGCG) demonstrated that different chemical shifts were observed for the phosphates of d(GpC) at positions 2 and 10 as well as for d(CpG) at positions 1, 3, 9, and 11 which indicates that the position in an oligonucleotide influences the chemical shift.[140] Additionally the more central the phosphate is, the more the signal appears at higher field. Exceptions at positions 3 and 9 were observed where signals were found at lower field strength than expected which suggests a break in conformation at these two locations. Discontinuities were also observed at these two positions (the sites of cleavage by the *Eco*RI nuclease) in the crystal structure as well which may mean that the solution and crystal structures have similar conformations. However of interest were certain differences between the crystal structure and the solution ^{31}P NMR which need to be explored in the future. Recently, additional research using the oxygen-17 labeling technique on d(GACGATATCGTC) has further confirmed and expanded work done with d(CGCGAATTCGCG). A reasonable correlation between ^{31}P NMR chemical shifts and the sum function of the base plane roll angles derived from Calladine's rule[143] was also shown to exist.

The synthesis of phosphonate analogs via the phosphite approach can proceed by two different pathways. The first involves conversion of **38a,b** to **39d** using alkyl halides in an Arbusov rearrangement.[91,129,144,145] The reaction, however, of support-bound internucleotide phosphites with alkyl iodides gave very low yields of the alkyl phosphonates.[91] An alternative approach where nucleoside trivalent alkyl phosphines or dinulceoside pentavalent alkyl phosphonates are condensed to the growing support-bound segment appears more attractive. Initially (Fig. 12) methyldichlorophosphine **(41a)** was condensed with **7** to form **42a,** which in turn reacted with **43** (R' = silica-based support covalently joined to the nucleoside through a succinate ester[35,36]) to yield **44** after iodine oxidation.[145] As expected from previous results with methoxydichlorophosphines[25] and as reported earlier by Engels for methyldichlorophosphines,[146] serious problems with this approach are the instability of **42a** toward hydrolysis and the generation of large amounts of 3' → 3' dinucleoside methyl phosphonates during synthesis of **42a.** These considerations led others to investigate nucleoside methyl phosphonamidites **(42b)** which were found to be much more stable and free of the 3' → 3' side products[62,147]; characteristics analogous to the nucleoside methylphosphoramidites.[72,73] Several comments regarding their synthesis and utilization should be addressed.

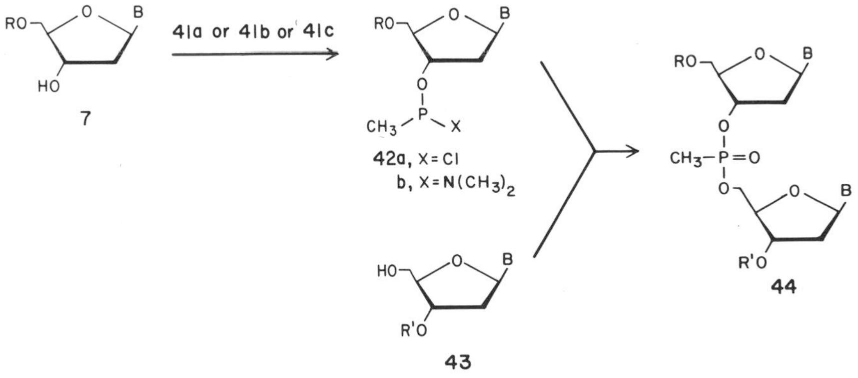

Fig. 12. Synthesis of oligonucleotides containing methylphosphonate internucleotide bonds. Abbreviations: R, trityl or dimethoxytrityl; R', silica, acetyl or benzoyl; *41a,* CH_3PCl_2; **41b**, $CH_3P(Cl)N(CH_3)_2$; **41c**, $CH_3P[N(CH_3)_2]_2$.

Compound **42b** can be prepared from **7** and either **41b** or preferably **41c**. The former synthesis method generates amine hydrochlorides which will destablize **42b** unless removed completely. The latter method involves activation of **41c**, a bis-N,N-dimethylaminomethylphosphine, with catalytic amounts of collidine hydrochloride and yields a stable intermediate (**42b**). The reaction is similar to the synthesis of nucleoside phosphoramidites from bis-N,N-dialkylaminomethoxyphosphines[88] and, by analogy, could probably be carried out more cleanly using dialkylammonium salts of various nitrogen heterocycles. Further activation and condensation with **43** (R' = benzoyl or acetyl) has been completed using imidazole,[62] tetrazole,[62,147] and 1H-benzotriazole.[147] Imidazole appears to be the reagent of choice as the side products observed with either tetrazole or extended treatment with 1H-benzotriazole are not produced. These side products appear to be due to acid-catalyzed ligand exchange to yield all possible diesters of **44** (3' → 3', 3' → 5', and 5' → 5'). Presumably imidazole being a weaker acid than the others does not catalyze this exchange reaction but is still sufficiently acidic to activate methyl phosphonamidites. In contrast, mononucleoside phosphoramidites containing methoxy rather than methyl on phosphorus in **42b** are not activated by imidazole but rather require tetrazole, the stronger acid.[62] Presumably the stronger acid is required to protonate the less basic phosphoramidites.

The diastereomers of **44** (R' = acetyl) can then be resolved by chromatography, converted to the 3'-phosphoromorpholidites, and condensed with **43** (R' = oligonucleotide attached to silica) to form chiral deoxyoligonucleotides.[91] In this way *lac* operator duplexes were synthesized so as to contain methyl phosphonates that were introduced stereospecifically at selected sites. Introduction of chiral phosphonates on the DNA face opposite the *lac* repressor recognition site had little effect on stability of the repressor–operator complex. When methyl phosphonates were introduced on the DNA face recognized by *lac* repressor, the affinity of repressor for the modified operators was dependent on the stereochemical configuration of the methyl phosphonate. The synthesis and biochemical properties of methyl phosphonate oligomers whose sequences are complementary to transfer RNA and to the Shine–Dalgarno sequence of 16 S rRNA have also been described.[148,149] These oligomers were shown to be capable of specifically inhibiting tRNA aminoacylation and protein synthesis both in cell-free systems and in living cells by virtue of their ability to bind to complementary sequences of cellular nucleic acid. All these results therefore suggest that alkyl phosphonates should be quite useful for studying certain biochemical reactions.

Recent experiments, especially by Miller *et al.*,[150] have shown that deoxyoligonucleotides having internucleotide phosphate triesters are resistant to nucleases and are taken up by cells. Because these oligomers hybridize effectively to unmodified polynucleotides, they are therefore candidates for antiviral drugs.[151] Since very efficient pathways now exist for synthesizing oligonucleotides using phosphite intermediates, this approach has been explored for preparing stable oligomers containing phosphate triesters (**39f**). Using **9a,** deoxyoligonucleotides containing phosphate triesters were synthesized on silica and removed from the support under conditions (concentrated ammonium hydroxide) where the triester is stable.[16] Since triesters are chiral, this method produces diastereomeric mixtures of products that increase in stereochemical complexity with each additional phosphate triester. Two methods are available for solving this problem.[132,133] One is to synthesize dinucleotide phosphate triesters in protected form and separate the diastereomers by silica gel column chromatography. Each diastereomer can then be converted to the dinucleotide 3′-phosphoramidite and used to synthesize optically pure deoxyoligonucleotides. The alternative method involves synthesis of a deoxyoligonucleotide containing diastereomers of phosphate triesters and then separating these diastereomers by HPLC. Using the latter approach to resolve diastereomers and two deoxynucleoside phosphoramidites (**28f** and **28f** where isopropoxy replaces methoxy as the phosphorus-protecting group), mixed triester–diester deoxyoligonucleotides corresponding to d(G-GAATTCC), a DNA segment recognized by *Eco*RI, were synthesized so as to contain an isopropyl group at the cleavage site of *Eco*RI (position 2 when phosphates are numbered left to right). In separate syntheses, phosphates at positions which have been proposed as backbone contact sites between enzyme and DNA (positions 1, 3, and 4) were also synthesized. It was found that none of the triester-containing octamers underwent detectable amounts of cleavage.[152] Unfortunately, negative controls at noncontact positions were not tested to determine if a DNA–protein complex can still form in the presence of triesters. This strategy could become quite useful for studying various biological processes or perhaps for inhibiting viral replication.

Deoxyoligonucleotides or oligoribonucleotides as the trichloroethyl[153,154] or 1,1,-dimethyl-2,2,2-trichloroethyl[15]-protected derivatives were synthesized and isolated as stable phosphites after chromatography. Of interest was the observation that dinucleoside phosphites in completely deprotected form were hydrolyzed by snake venom phosphodiesterase but not ribonuclease A or spleen phosphodiesterases.[154] As these compounds are electrically neutral and partially resistant to nucleases, the possibility exists that they can be taken up by cells and therefore may be potential candidates as antiviral agents.

Phosphoramidate internucleotide linkages can also be synthesized from trivalent phosphites. A general approach involves conversion of **38b** to **39e** using iodine and a primary amine which is followed by removal of the trichloroethyl group to yield **40e**. These reactions are reported to be essentially quantitative and very rapid.[129] An alternative series of phosphoramidates can also be generated by first condensing *N,N*-diethyldichlorophosphine with protected ribonucleotides to yield **38c**. This compound can then be oxidized to yield the thio, seleno, and imino analogs of the dinucleoside phosphoramidate.[155] Since all these syntheses were completed using only uridine, the applicability of the procedures to the other nucleosides and deoxynucleosides must still be investigated. The synthesis of a dinucleotide where the phosphoramidate joins a nucleotide to phosphorus through a phosphorus–nitrogen bond can most easily be completed as well using trivalent phosphites (Fig. 13). Compound **45** reacts rapidly (2 min) with 5′-azido-5′-deoxythymidine **(46)** to yield **47** in good yield.[156] Of interest was that selective elimination of an ethyl group over nucleoside from the mixed phosphite was enhanced by the addition of an excess of lithium chloride. In the absence of lithium chloride, the major product was deoxythymidine 5′-diethylphosphoramidate. This method has now been used to synthesize **50,** a molecule that intercalates into complementary deoxypolynucleotides (Fig. 14).[157] Synthesis of **49** proceeds by coupling **7** and **48** using *o*-chlorophenyldichlorophosphite. Since *o*-chlorophenoxy is the preferred leaving group, reaction of **49** with ethyl azidoacetate leads to the phosphoramidate. Further condensation, first with 1,4-diaminobutane and then the *p*-nitrophenyl ester of 3,7-diamino-6-*p*-carboxyphenyl-5-methylphenanthridinium iodide followed by

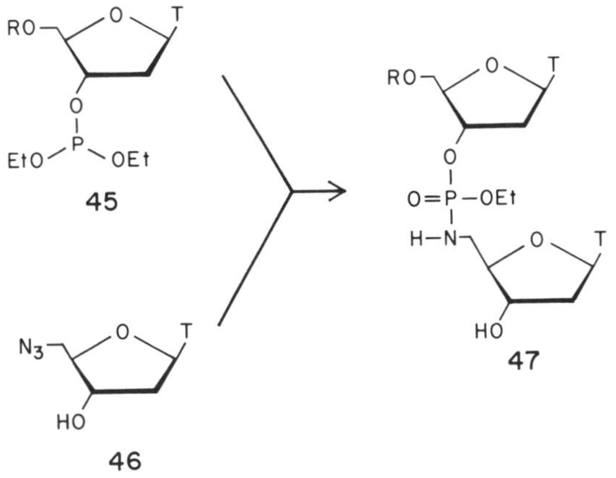

Fig. 13. Synthesis of a dinucleotide containing a phosphoramidate internucleotide bond. Et, Ethyl.

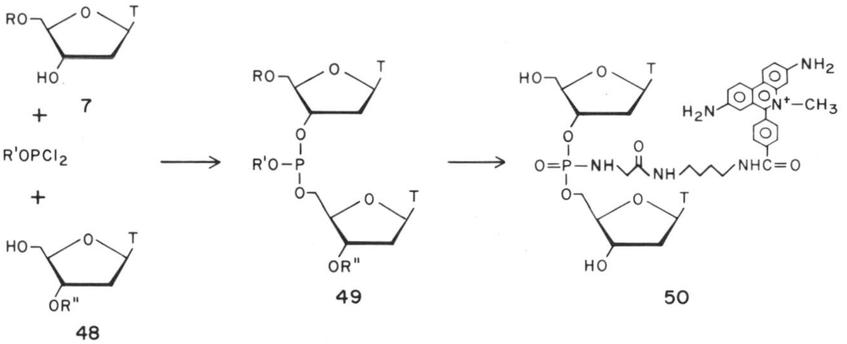

Fig. 14. Synthesis of a DNA-binding dinucleotide. Abbreviations: R, phenoxyacetyl; R', o-chlorophenyl; R", monomethoxytrityl.

acid cleavage of the 3'-protecting group leads to **50**. This compound was found to form a stable complex with poly(A) due to both the formation of Watson–Crick base pairs and the interaction of the phenanthridinium ring with adenine bases. The synthesis of **50** illustrates a general pathway for tagging polynucleotides via phosphite intermediates with various reporter or chemically reactive groups. For example, this chemistry can probably be used to label polynucleotides with electron spin resonance or fluorescence probes and with chemical cross-linking agents for studying various interactions of polynucleotides or between polynucleotides and proteins.

In addition to phosphorus, several base and sugar analogs have been prepared for various biochemical studies. One particularly interesting class involves oligonucleotides which contain aliphatic amino[158] or thiol[159] groups at the 5'-terminus. Fluorescent chromophores have been attached to these functional groups and, in one case, used to develop a nonradioactive methodology for sequencing DNA by the dideoxy procedure.[158] The same concept has also been used for attachment of biotin to the 5'-hydroxyl of an oligonucleotide through an amino alkyl phosphoramidate[160] or phosphate[161] linkage. These 5'-biotin-labeled oligonucleotides, when hybridized to target plasmid DNA immobilized on nitrocellulose filters, could detect 2 fmole of target.[160] Through the synthesis of deoxyoligonucleotides containing modified bases, new insights have been gained regarding the mechanisms whereby proteins recognize and interact with nucleic acids.[162,163] For example, by using deoxyuridine, 5-bromodeoxyuridine, 5-methyldeoxycytidine, and 5-bromodeoxycytidine to replace deoxythymidine at specific sites in the *lac* operator, the *lac* repressor was shown to recognize the 5-methyl group on thymine.[164,165] More recently this same approach has been used to demonstrate that certain thymine 5-

methyl groups on thymine are also recognized by *Escherichia coli* RNA polymerase,[63,84] the lambda *c*I repressor,[84] and the *Eco*RII restriction endonuclease.[166] Other research has shown that the 2-amino group on guanine is recognized by the *lac* repressor.[167] These results have prompted the development of additional analogs such as 7-deazadeoxyguanosine[168] and N^6-methyldeoxyadenosine[169] which have now been incorporated into deoxyoligonucleotides. The 7-deazadeoxyguanosine analog should prove quite useful as a probe for protein recognition of the N-7 position on guanine. Another very recent development is the use of synthetic probes containing deoxyinosine at ambiguous codon positions.[170,171] The analog has been proposed as an inert base that neither destablizes nor contributes at mismatched sites toward formation of a DNA duplex. Its utility has also been demonstrated in the isolation of the human cholecystokinin gene.[172]

VI. CONCLUDING REMARKS

The development of DNA and RNA synthesis methodologies that can be used by molecular biologists, biologists, and chemists is undoubtedly the most significant advance for some time in nucleic acid chemistry. When combined with other developments in molecular cloning and nucleic acid sequencing techniques, these advances have had a profound effect on many biological studies. Synthetic DNA is now used routinely to isolate and synthesize genes, engineer proteins, prime the sequencing of DNA, study the regulation and control of gene expression, and probe the structure of DNA and RNA. For the future, many infectious diseases and genetic disorders will probably be diagnosed using synthetic DNA as well. Clearly as a result of all these advances in DNA technology, for the first time, the research scientist has total control over any DNA sequence being investigated. This flexibility will undoubtedly lead to major research and technological advances in the years to come.

ACKNOWLEDGMENTS

The research cited from the author's laboratory was supported by grants from the National Science Foundation (PCM77-20618) and the National Institutes of Health (GM21120 and GM25680). Several excellent and imaginative graduate and postdoctoral students contributed immeasurably to the development of the chemical methodology reported. Their names appear in the appropriate references. I also wish to thank A. Sirimarco for help in preparing this contribution.

REFERENCES

1. Michelson, A. M., and Todd, A. R. (1955). *J. Chem. Soc., Part 3,* pp. 2632–2638.
2. Khorana, H. G., Büchi, H., Ghosh, H., Gupta, N., Jacob, T. M., Kössel, H., Morgan, A. R., Narang, S. A., Ohtsuka, E., and Wells, R. D. (1966). *Cold Spring Harbor Symp. Quant. Biol.* **31,** 39–49.
3. Khorana, H. G., Agarwal, K. L., Büchi, H., Caruthers, M. H., Gupta, N. K., Kleppe, K., Kumar, A., Ohtsuka, E., RajBhandary, U. L., van de Sande, J. H., Sgaramella, V., Terao, T., Weber, H., and Yamada, T. (1972). *J. Mol. Biol.* **72,** 209–217.
4. Khorana, H. G., Agarwal, K. L., Besmer, P., Büchi, H., Caruthers, M. H., Cashion, P. J., Fridkin, M., Jay, E., Kleppe, K., Kleppe, R., Kumar, A., Loewen, P. C., Miller, R. C., Minamoto, K., Panet, A., RajBhandary, U. L., Ramamoorthy, B., Sekiya, T., Takeya, T., and van de Sande, J. H. (1976). *J. Biol. Chem.* **251,** 565–570.
5. Letsinger, R. L., and Mahadevan, V. (1965). *J. Am. Chem. Soc.* **87,** 3526–3527.
6. Letsinger, R. L., and Mahadevan, V. (1966). *J. Am. Chem. Soc.* **88,** 5319–5324.
7. Letsinger, R. L., Caruthers, M. H., Miller, P. S., and Ogilvie, K. K. (1967). *J. Am. Chem. Soc.* **89,** 7146–7147.
8. Letsinger, R. L., and Ogilvie, K. K. (1969). *J. Am. Chem. Soc.* **91,** 3350–3355.
9. Letsinger, R. L., Ogilvie, K. K., and Miller, P. S. (1969). *J. Am. Chem. Soc.* **91,** 3360–3365.
10. Letsinger, R. L., Finnan, J. L., Heavner, G. A., and Lunsford, W. B. (1975). *J. Am. Chem. Soc.* **97,** 3278–3279.
11. Letsinger, R. L., and Lunsford, W. B. (1976). *J. Am. Chem. Soc.* **98,** 3655–3661.
12. Itakura, K., Rossi, J., and Wallace, R. B. (1984). *Annu. Rev. Biochem.* **53,** 323–356.
13. Narang, S. (1983). *Tetrahedron* **39,** 3–22.
14. Ohtsuka, E., Ikehara, M., and Söll, D. (1982). *Nucleic Acids Res.* **10,** 6553–6570.
15. Letsinger, R. L., Groody, E. P., and Tanaka, T. (1982). *J. Am. Chem. Soc.* **104,** 6805–6806.
16. Letsinger, R. L., Groody, E. P., Lander, N., and Tanaka, T. (1984). *Tetrahedron* **40,** 137–143.
17. Ogilvie, K. K., Theriault, N., and Sadana, K. L. (1977). *J. Am. Chem. Soc.* **99,** 7741–7743.
18. Ogilvie, K. K., and Theriault, N. Y. (1979). *Tetrahedron Lett.,* pp. 2111–2114.
19. Ogilvie, K. K., Schifman, A. L., and Penny, C. L. (1979). *Can. J. Chem.* **57,** 2230–2238.
20. Ogilvie, K. K., and Theriault, N. Y. (1979). *Can. J. Chem.* **57,** 3140–3144.
21. Ogilvie, K. K., and Nemer, M. J. (1980). *Can. J. Chem.* **58,** 1389–1397.
22. Ogilvie, K. K., Nemer, M. J., Theriault, N., Pon, R., and Siefert, J. (1980). *Nucleic Acids Symp. Ser.* **7,** 147–150.
23. Ogilvie, K. K., and Pon, R. T. (1980). *Nucleic Acids Res.* **8,** 2105–2115.
24. Daub, G. W., and van Tamelen, E. E. (1977). *J. Am. Chem. Soc.* **99,** 3526–3528.
25. Matteucci, M. D., and Caruthers, M. H. (1981). *J. Am. Chem. Soc.* **103,** 3185–3191.
26. Ogilvie, K. K., Theriault, N. Y., Seifert, J.-M., Pon, R. T., and Nemer, M. J. (1980). *Can. J. Chem.* **58,** 2686–2693.
27. Gait, M. J., and Sheppard, R. C. (1977). *Nucleic Acids Res.* **4,** 4391–4410.
28. Duckworth, M. L., Gait, M. J., Goelet, P., Hond, G. F., Singh, M., and Titmas, R. C. (1981). *Nucleic Acids Res.* **7,** 1691–1706.
29. Narang, C. K., Brunfeldt, K., and Norris, K. E. (1977). *Tetrahedron Lett.* **18,** 1819–1822.

30. Miyoshi, K., Miyake, T., Hozumi, T., and Itakura, K. (1980). *Nucleic Acids Res.* **8,** 5473–5489.
31. Miyoshi, K., Arentzen, R., Huang, T., and Itakura, K. (1980). *Nucleic Acids Res.* **8,** 5507–5517.
32. Ito, H., Ike, Y., Ikuta, S., and Itakura, K. (1982). *Nucleic Acids Res.* **10,** 1755–1769.
33. Potapov, V. K., Veiko, V. P., Koroleva, O., and Shabarova, Z. (1979). *Nucleic Acids Res.* **6,** 2041–2056.
34. Crea, R., and Horn, T. (1980). *Nucleic Acids Res.* **8,** 2331–2348.
35. Caruthers, M. H., Beaucage, S. L., Efcavitch, J. W., Fisher, E. F., Matteucci, M. D., and Stabinsky, Y. (1980). *Nucleic Acids Symp. Ser.* **7,** 215–223.
36. Matteucci, M. D., and Caruthers, M. H. (1980). *Tetrahedron Lett.* **21,** 719–722.
37. Köster, H. (1972). *Tetrahedron Lett.* **13,** 1527–1530.
38. Gough, G. R., Brunden, M. J., and Gilham, P. T. (1981). *Tetrahedron Lett.* **22,** 4177–4180.
39. Efimov, V. A., Reverdatto, S. V., and Chakhmakhcheva, O. G. (1982). *Nucleic Acids Res.* **10,** 6675–6694.
40. Adams, S. P., Kavka, K. S., Wykes, E. J., Holder, S. B., and Galluppi, G. R. (1983). *J. Am. Chem. Soc.* **105,** 661–663.
41. Kohli, V. P., Balland, A., Wintzerith, M., Sauerwald, R., Staub, A., and Lecocq, J. P. (1982). *Nucleic Acids Res.* **10,** 7439–7448.
42. Köster, H., Biernat, J., McManus, J., Wolter, A., Stumpe, A., Narang, C. K., and Sinha, H. D. (1984). *Tetrahedron* **40,** 103–112.
43. Caruthers, M. H., Beaucage, S. L., Becker, C., Efcavitch, W., Fisher, E. F., Galluppi, G., Goldman, R., deHaseth, P., Martin, F., Matteucci, M., and Stabinsky, Y. (1982). *In* "Genetic Engineering, Principles and Methods" (J. K. Setlow and A. Hollaender, eds.), Vol. 4, pp. 1–17. Plenum, New York.
44. Caruthers, M. H. (1983). *In* "Methods of DNA and RNA Sequencing" (S. M. Weissman, ed.), pp. 1–22. Praeger, New York.
45. Caruthers, M. H. (1981). *In* "Recombinant DNA, Proceedings of the Third Cleveland Symposium on Macromolecules" (A. G. Walton, ed.), pp. 261–272. Elsevier, The Netherlands.
46. Chow, F., Kempe, T., and Palm, G. (1981). *Nucleic Acids Res.* **9,** 2807–2817.
47. Caruthers, M. H., Stabinsky, Y., Stabinsky, Z., and Peters, M. (1982). *In* "Promoters: Structure and Function" (R. L. Rodriquez and M. J. Chamberlin, eds.), pp. 432–451. Praeger, New York.
48. Waddell, T. G., Leyden, D. E., and DeBello, M. T. (1981). *J. Am. Chem. Soc.* **103,** 5303–5307.
49. Gough, G. R., Brunden, M. J., and Gilham, P. T. (1983). *Tetrahedron Lett.* **24,** 5321–5324.
50. Alvarado-Urbina, G., Sathe, G. M., Liu, W. C., Gillen, M. F., Duck, P. D., Bender, R., and Ogilvie, K. K. (1981). *Science* **214,** 270–274.
51. Tanaka, T., and Letsinger, R. L. (1982). *Nucleic Acids Res.* **10,** 3249–3260.
52. Fisher, E. F., Feist, P. L., Beaucage, S. L., Meyers, R. M., Tjian, R., and Caruthers, M. H. (1984). *Biochemistry* **23,** 5938–5944.
53. Weinmann, G., and Khorana, H. G. (1962). *J. Am. Chem. Soc.* **84,** 419–430.
54. Letsinger, R. L., Caruthers, M. H., and Jerina, D. M. (1967). *Biochemistry* **6,** 1379–1388.
55. Jayaraman, K., and McClaugherty, H. (1982). *Tetrahedron Lett.* **23,** 5377–5380.
56. Ogilvie, K. K., and Nemer, M. J. (1981). *Tetrahedron Lett.* **22,** 2531–2534.

57. Matteucci, M. D. (1980). Ph.D. Thesis, Department of Chemistry and Biochemistry, University of Colorado, Boulder.
58. Kohli, V., Blocker, H., and Koster, H. (1980). *Tetrahedron Lett.* **21**, 2683–2686.
59. Matteucci, M. D., and Caruthers, M. H. (1980). *Tetrahedron Lett.* **21**, 3243–3246.
60. Kierzek, R., Ito, H., Batt, R., and Itakura, K. (1981). *Tetrahedron Lett.* **22**, 3761–3764.
61. Ito, H., Ike, Y., Ikuta, S., and Itakura, K. (1982). *Nucleic Acids Res.* **10**, 1755–1769.
62. Dorman, M. A., Noble, S. A., McBride, L. J., and Caruthers, M. H. (1984). *Tetrahedron* **40**, 95–102.
63. Caruthers, M. H., Beaucage, S. L., Efcavitch, J. W., Fisher, E. F., Goldman, R. A., deHaseth, P. L., Mandecki, W., Matteucci, M. D., Rosendahl, M. S., and Stabinsky, Y. (1983). *Cold Spring Harbor Symp. Quant. Biol.* **47**, 411–418.
64. deHaseth, P. L., Goldman, R. A., Cech, C. L., and Caruthers, M. H. (1983). *Nucleic Acids Res.* **11**, 773–787.
65. Gottlieb, P., Nasoff, M. S., Fisher, E. F., Walsh, A. M., and Caruthers, M. H. (1985). *Nucleic Acids Res.* **13**, 6621–6634.
66. Kume, A., Iwase, R., Sekine, M., and Hata, T. (1984). *Nucleic Acids Res.* **12**, 8525–8538.
67. McBride, L. J., and Caruthers, M. H. (1983). *Tetrahedron Lett.* **24**, 2953–2956.
68. Froehler, B. C., and Matteucci, M. D. (1983). *Nucleic Acids Res.* **11**, 8031–8036.
69. Caruthers, M. H., McBride, L. J., Bracco, L. P., and Dubendorff, J. W. (1985). *Nucleosides Nucleotides* **4**, 95–105.
70. Fourrey, J. L., and Shire, D. J. (1981). *Tetrahedron Lett.* **22**, 729–732.
71. Cao, T. M., Bingham, S. E., and Sung, M. T. (1983). *Tetrahedron Lett.* **24**, 1019–1020.
72. Beaucage, S. L., and Caruthers, M. H. (1981). *Tetrahedron Lett.* **22**, 1859–1862.
73. McBride, L. J., and Caruthers, M. H. (1983). *Tetrahedron Lett.* **24**, 245–248.
74. Seliger, H. and Gupta, K. C. (1985). *Nucleosides Nucleotides* **4**, 249.
75. Schwarz, M. W., and Pfleiderer, W. (1984). *Tetrahedron Lett.* **25**, 5513–5516.
76. Fourrey, J. L., and Varenne, J. (1983). *Tetrahedron Lett.* **24**, 1963–1966.
77. Sinha, N. D., Biernat, J., and Köster, H. (1983). *Tetrahedron Lett.* **24**, 5843–5846.
78. Claesen, C., Tesser, G. I., Dreif, C. E., Marugg, J. E., van der Marel, G. A., and van Boom, J. H. (1984). *Tetrahedron Lett.* **25**, 1307–1310.
79. Beiter, A. H., and Pfleiderer, W. (1984). *Tetrahedron Lett.* **25**, 1975–1978.
80. Hering, G., Stöcklein-Schneirderwind, R., Ugi, I., Pathak, T., Balgobin, N., and Chattopadhyaya, J. (1985). *Nucleosides Nucleotides* **4**, 169–171.
81. Marugg, J. E., Dreef, C. E., van der Marel, G. A., and van Boom, J. H. (1984). *Recl. Trav. Chim. Pays-Bas* **103**, 97–98.
82. Dorper, T., and Winnacker, E. L. (1983). *Nucleic Acids Res.* **11**, 2575–2584.
83. Froehler, B. C., and Matteucci, M. D. (1983). *Tetrahedron Lett.* **24**, 3171–3174.
84. Caruthers, M. H. (1985). *Science* **230**, 281–285.
85. Gao, X., Gaffney, B. L., Senior, M., Riddle, R. R., and Jones, R. A. (1985). *Nucleic Acids Res.* **13**, 573–584.
86. Caruthers, M. H. (1982). *In* "Chemical and Enzymatic Synthesis of Gene Fragments: A Laboratory Manual" (H. G. Gassen and A. Lang, eds.), pp. 71–79. Verlag-Chemie, Weinheim.
87. Beaucage, S. L. (1984). *Tetrahedron Lett.* **25**, 375–378.
88. Barone, A. D., Tang, J.-Y., and Caruthers, M. H. (1984). *Nucleic Acids Res.* **12**, 4051–4061.
89. Efcavitch, J. W., and Heiner, C. (1985). *Nucleosides Nucleotides* **4**, 267.
90. Kumar, G., and Poonian, M. S. (1984). *J. Org. Chem.* **49**, 4905–4912.

91. Noble, S. A., Fisher, E. F., and Caruthers, M. H. (1984). *Nucleic Acids Res.* **12**, 3387–3404.
92. Caruthers, M. H. (1982). *In* "From Gene to Protein: Translation into Biotechnology" (F. Ahmad, J. Schultz, W. J. Whelan, and E. E. Smith, eds.), pp. 235–247. Academic Press, New York.
93. Reese, C. B., and Ubasawa, A. (1980). *Tetrahedron Lett.* **21**, 2265–2268.
94. Gaffney, B. L., and Jones, R. A. (1982). *Tetrahedron Lett.* **23**, 2253–2256.
95. Adamiak, R. W., and Biala, E. (1985). *Nucleic Acids Res.* **13**, 2989–3003.
96. Daskalov, H. P., Sekine, M., and Hata, T. (1981). *Bull. Chem. Soc. Jpn.* **54**, 3076–3085.
97. Sung, W. L. (1982). *J. Org. Chem.* **47**, 3623–3628.
98. Sung, W. L., and Narang, S. A. (1982). *Can. J. Chem.* **60**, 111–120.
99. Reese, C. B., and Richards, K. H. (1985). *Tetrahedron Lett.* **26**, 2245–2248.
100. Ohtsuka, E., Yamane, A., and Ikehara, M. (1983). *Nucleic Acids Res.* **11**, 1325–1335.
101. Pon, R. T., Damha, M. J., and Ogilvie, K. K. (1985). *Tetrahedron Lett.* **26**, 2525–2528.
102. Pon, R. T., Damha, M. J., and Ogilvie, K. K. (1985). *Nucleic Acids Res.* **13**, 6447–6465.
103. Fisher, E. F., and Caruthers, M. H. (1983). *Nucleic Acids Res.* **11**, 1589–1599.
104. Gait, M. J., Matthes, H. W. D., Singh, M., Sproat, B., and Titmas, R. G. (1982). *In* "Chemical and Enzymatic Synthesis of Gene Fragments" (H. G. Gassen and A. Lang, eds.), pp. 1–42. Verlag Chemie, Weinheim.
105. Sanchez-Pescador, R., and Urdea, M. S. (1984). *DNA* **3**, 339–343.
106. Frank, R., Heikens, W., Heisterberg-Montsis, G., and Blöcker, H. (1983). *Nucleic Acids Res.* **11**, 4365–4377.
107. Ott, J., and Eckstein, F. (1984). *Nucleic Acids Res.* **12**, 9137–9142.
108. Hunkapillar, M., Kent, S., Caruthers, M., Dreyer, W., Firca, J., Giffin, C., Horvath, S., Hunkapillar, T., Tempst, P., and Hood, L. (1984). *Nature (London)* **310**, 105–111.
109. Reese, C. B. (1978). *Tetrahedron* **34**, 3143–3179.
110. Maxam, A. M., and Gilbert, W. (1980). *In* "Methods in Enzymology" (L. Grossman and K. Moldave, eds.), Vol. 65, pp. 499–560. Academic Press, New York.
111. Gait, M. J., Matthes, H. W. P., Singh, M., Sproat, B. S., and Titmas, R. C. (1982). *Nucleic Acids Res.* **10**, 1755–1769.
112. Markham, A. F., Edge, M. D., Atkinson, T. C., Greene, A. R., Heathcliffe, G. R., Newton, C. R., and Scanlon, D. (1980). *Nucleic Acids Res.* **8**, 5193–5205.
113. Sanger, F., Donelson, A. R., Coulson, H., Kössel, H., and Fisher, D. (1973). *Proc. Natl. Acad. Sci. U.S.A.* **70**, 1209–1213.
114. Jay, E., Bambara, P., and Wu, R. (1974). *Nucleic Acids Res.* **1**, 331–356.
115. Rushlow, K. (1983). *Focus (Bethesda Research Lab., Bethesda, Md.)* **5**, 1–4.
116. Ogilvie, K. K., and Nemer, M. J. (1980). *Tetrahedron Lett.* **21**, 4159–4162.
117. Pon, R. T., and Ogilvie, K. K. (1984). *Tetrahedron Lett.* **25**, 713–716.
118. Pon, R. T., and Ogilvie, K. K. (1984). *Nucleosides Nucleotides* **3**, 485–498.
119. Ogilvie, K. K., Nemer, M. J., and Gillen, M. F. (1984). *Tetrahedron Lett.* **25**, 1669–1672.
120. Kempe, T., Chow, F., Sundquist, W. I., Nordi, T. J., Paulson, B., and Peterson, S. M. (1982). *Nucleic Acids Res.* **10**, 6695–6714.
121. Seliger, H., Zeh, D., and Azuru, G. (1983). *Chem. Scr.* **22**, 95–101.
122. Usman, N., Pon, R. T., and Ogilvie, K. K. (1985). *Tetrahedron Lett.* **26**, 4567–4570.
123. Caruthers, M. H., Dellinger, D., Prosser, K., Barone, A. D., Dubendorff, J. W., Kierzek, R., and Rosendahl, M. (1986). *Chem. Scr.* **26**, 25–30.
124. Jones, S. S., and Reese, C. B. (1979). *J. Chem. Soc., Perkin Trans. 1*, pp. 2762–2765.
125. Kohler, W., Schlosser, W., Charubala, G., and Pfleiderer, W. (1978). *In* "Chemistry

and Biology of Nucleosides and Nucleotides'' (R. E. Harmon, R. K. Robin S, and L. B. Townsend, eds.), pp. 347–358. Academic Press, New York.
126. Ogilvie, K. K., and Entwistle, D. W. (1981). *Carbohydr. Res.* **89**, 203–210.
127. Markiewicz, W., Biala, T., and Kierzek, R. (1984). *Bull. Pol. Acad. Sci. Chem.* **32**, No. 11–12, 433–451.
128. Eckstein, F. (1967). *Tetrahedron Lett.* **7**, 1157–1160.
129. Nemer, M. J., and Ogilvie, K. K. (1980). *Tetrahedron Lett.* **21**, 4149–4152.
130. Connolly, B. A., Potter, V. L., Eckstein, F., Pingond, A., and Grotjahn, L. (1984). *Biochemistry* **23**, 3443–3453.
131. Cosstick, R., and Eckstein, F. (1985). *Biochemistry* **24**, 3630–3638.
132. Stec, W. J., Zon, G., Egan, W., and Stec, B. (1984). *J. Am. Chem. Soc.* **106**, 6077–6079.
133. Stec, W. J., and Zon, G. (1984). *Tetrahedron Lett.* **25**, 5275–5278.
134. Eckstein, F. (1983). *Angew. Chem., Int. Ed. Eng.* **22**, 423–439.
135. Petersheim, M., Mehdi, S., and Gerlt, J. A. (1984). *J. Am. Chem. Soc.* **106**, 439–440.
136. Shah, D. O., Lai, K., and Gorenstein, D. G. (1984). *J. Am. Chem. Soc.* **106**, 4302–4303.
137. Gorenstein, D. G., Lai, K., and Shah, D. O. (1984). *Biochemistry* **23**, 6717–6723.
138. Connolly, B. A., and Eckstein, F. (1984). *Biochemistry* **23**, 5523–5527.
139. Ott, J., and Eckstein, F. (1985). *Nucleic Acids Res.* **13**, 6317–6330.
140. Ott, J., and Eckstein, F. (1983). *J. Am. Chem. Soc.* **105**, 5879–5886.
141. Seela, F., Ott, J., and Potter, B. V. L. (1983). *J. Am. Chem. Soc.* **105**, 5879–5886.
142. Potter, B. V. L., and Eckstein, F. (1983). *Nucleic Acids Res.* **11**, 7087–7103.
143. Calladine, C. R. (1982). *J. Mol. Biol.* **161**, 343–352.
144. Stec, W. J., Zon, G., Egan, W., Byrd, A. R., Phillips, R., and Gallo, K. A. (1985). *J. Org. Chem.* **50**, 3908–3913.
145. Sinha, N. D., Grossbruchhaus, V., and Koster, H. (1983). *Tetrahedron Lett.* **24**, 877–880.
146. Engels, J., and Jäger, A. (1982). *Angew. Chem., Int. Ed. Engl.* **21**, 912.
147. Jäger, A., and Engels, J. (1984). *Tetrahedron Lett.* **25**, 1437–1440.
148. Miller, P. S., McParland, K. B., Jayaraman, K., and Ts'o, P. O. P. (1981). *Biochemistry* **20**, 1874–1880.
149. Jayaraman, K., McParland, K. P., Miller, P. S., and Ts'o, P. O. P. (1981). *Proc. Natl. Acad. Sci. U.S.A.* **78**, 1537–1541.
150. Miller, P. S., Chandrasegaran, S., Dru, D., Pulford, S. W., and Kan, L. S. (1982). *Biochemistry* **21**, 5468–5474, and references cited therein.
151. Zemecnik, P. C., and Stephenson, M. L. (1978). *Proc. Natl. Acad. Sci. U.S.A.* **75**, 2880–284.
152. Stec, W. J., Zon, G., Gallo, K. A., and Byrd, R. A. (1985). *Tetrahedron Lett.* **26**, 2191–2194.
153. Melnick, B. P., Finnan, J. L., and Letsinger, R. L. (1980). *J. Org. Chem.* **45**, 2715–2716.
154. Ogilvie, K. K., and Nemer, M. J. (1980). *Tetrahedron Lett.* **21**, 4145–4148.
155. Nemer, M. J., and Ogilvie, K. K. (1980). *Tetrahedron Lett.* **21**, 4153–4154.
156. Letsinger, R. L., and Heavner, G. A. (1975). *Tetrahedron Lett.* **16**, 147–150.
157. Letsinger, R. L., and Schott, M. E. (1981). *J. Am. Chem. Soc.* **103**, 7394–7396.
158. Smith, L. M., Fung, S., Hunkapillar, M. W., Hunkapillar, T. J., and Hood, L. E. (1985). *Nucleic Acids Res.* **13**, 2399–2412.
159. Connolly, B. A. (1985). *Nucleic Acids Res.* **13**, 4485–4502.
160. Chollet, A., and Kawashima, E. (1985). *Nucleic Acids Res.* **13**, 1529–1541.

161. Kempe, T., Sundquist, W. I., Chow, F., and Hu, S.-L. (1985). *Nucleic Acids Res.* **13,** 45–57.
162. Caruthers, M. H. (1980). *Acc. Chem. Res.* **13,** 155–160.
163. Modrich, P. (1982). *CRC Crit. Rev. Biochem.* **13,** 287–323.
164. Goeddel, D. V., Yansura, D. G., and Caruthers, M. H. (1977). *Nucleic Acids Res.* **4,** 3039–3054.
165. Fisher, E. F., and Caruthers, M. H. (1979). *Nucleic Acids Res.* **7,** 401–416.
166. Yolov, A. A., Vinogradova, M. N., Gromova, E. S., Rosenthal, A., Cech, D., Veiko, V. P., Metelev, V. G., Kosykh, V. G., Buryanov, Ya. I., Bayev, A. A., and Shabarova, Z. A. (1985). *Nucleic Acids Res.* **13,** 8983–8998.
167. Yansura, D. G., Goeddel, D. V., Kundu, A., and Caruthers, M. H. (1979). *J. Mol. Biol.* **133,** 117–135.
168. Seela, F., and Driller, H. (1985). *Nucleic Acids Res.* **13,** 911–926.
169. Delort, A. M., Guy, A., Molko, D., and Téoule, R. (1985). *Nucleosides Nucleotides* **4,** 201–203.
170. Takahashi, Y., Kato, K., Hayashizaki, Y., Wakabayashi, T., Ohtsuka, E., Matsuki, S., Ikehara, M., and Matsubara, K. (1985). *Proc. Natl. Acad. Sci. U.S.A.* **82,** 1931–1935.
171. Martin, F. H., and Castro, M. M. (1985). *Nucleic Acids Res.* **24,** 8927–8938.
172. Ohtsuka, E., Matsuki, S., Ikehara, M., Takahashi, Y., and Matsubara, K. (1985). *J. Biol. Chem.* **260,** 2605–2608.

4

Synthetic Gene Assembly, Cloning, and Expression

ROLAND BROUSSEAU
Montreal Biotechnology Research Institute
Montreal, Quebec, Canada H4C 2K3

RAY WU
Department of Biochemistry
Section of Molecular and Cell Biology
Cornell University
Ithaca, New York 14853

WING SUNG
Division of Biological Sciences
National Research Council of Canada
Ottawa, Ontario, Canada K1A OR6

SARAN A. NARANG
Division of Biological Sciences
National Research Council of Canada
Ottawa, Ontario, Canada K1A OR6

I. INTRODUCTION

Chromosomes control the hereditary properties of all living entities and are linear or circular collections of specific factors called genes. Each gene can affect the character of an organism in a highly specific way; from the humblest virus to the music composer, we are what our genes make us, the only difference being in the complexity of the message. Only 3200 bases of DNA are necessary to encode for hepatitis B virus; the molecular geneticist's workhorse, *Escherichia coli,* contains a single DNA molecule made up of approximately 4 million base pairs; for a human being about a thousand times more are needed. Deoxyribonucleic acid carries messages in two tracks, rather like a tape recording in which there are specific instructions for a job to be done. Exact copies can be made from it as from a recorded tape, so that this information can be used again and also transferred to the next generation of cells. Recorded on this tape is the

95

Synthesis and Applications
of DNA and RNA

genetic language written in the universal code that tells our history of evolution. There are only four letters in the alphabet of this language—A, G, T, and C. In permutations of the order of these four letters lie instructions in terms of space and time that will determine our physical characteristics and probably a fair bit of our psychological ones as well.

The most straightforward way to decode and understand these messages has been to develop the methodology of the synthesis of these DNA sequences. Such tailor-made DNA is of great importance in our understanding of the molecular biology of genes. The elucidation of the genetic code using synthetic DNA containing repeating base sequences is the first example of such achievements. Thanks to intensive work over the past 10 years, there has been considerable progress in the methods for gene synthesis, so that chemically synthesizing a gene has now become a well-accepted method of obtaining it in a form suitable for expression. While the labor involved is still considerable, the flexibility and convenience of having complete control over restriction sites, codon usage, and initiation and termination sites leads more and more people to this approach, even for large genes. This is especially so since the availability of commercial solid-phase DNA synthesizers which handle the tedious and repetitive chemical coupling steps needed to synthesize oligonucleotides.

In this chapter we will review the work done on chemical–enzymatic synthesis of expressed genes, with the exclusion of the very well-known work of Khorana et al. (1976) on tRNA genes and work done on synthetic DNA control regions such as the lac operator (Marians et al., 1976; Heyneker et al., 1976). Also we will not include work such as that of Goeddel et al. (1979b) in which a synthetic part is added to a natural gene.

II. GENERAL METHODOLOGY

Nucleic acids are often of high molecular weight; it is therefore advisable or even necessary to couple methods of organic chemistry with biochemical techniques to prepare DNAs of known sequence. Such a strategy will also favor the selection of biologically active molecules. Although it has become easy to synthesize chemically a sequence 100–200 bases long which may be for the most part chemically correct, it may not be biologically active. The use of enzymes and recombinant DNA technology discriminates against degraded or partially blocked molecules and selects only those which are biologically viable and useful.

A. Assembly of the DNA Pieces

1. Complete Chemical Synthesis of Both Strands of the DNA Molecule

DNA-joining enzymes identified as DNA ligases join together DNA chains by transmutating the high-energy pyrophosphate linkage of a nucleotide cofactor into a phosphodiester bond between 5'-phosphoryl and 3'-hydroxyl termini. The termini of strands to be joined by ligase must be, in general, abutted by base-pairing to a complementary strand; this ensures proper alignment and preservation of the nucleotide sequence of DNA (Fig. 1). Such enzymes play an essential and determinant role in synthetic DNA assembly.

The assembly of a synthetic gene basically involves the chemical synthesis of a number of complementary oligodeoxyribonucleotides, followed by their ligation together with a suitable ligase enzyme, T4 DNA ligase. The basic approach has been established (Agarwal *et al.,* 1970; Khorana *et al.,* 1976) in Khorana's group as far back as 1970 and has not been fundamentally modified since; however, the speed and ease of the chemical operations have improved dramatically, as described elsewhere in this book.

We have presented several recent gene syntheses in Table I, in Section III of this chapter, where the reader will find references to the details of synthesis and vector contruction. For those who only wish a brief exposition on the state of the field, a good example of recent work is given by

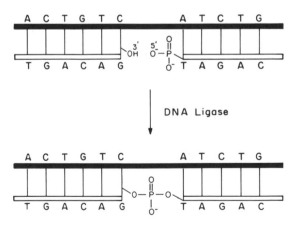

Fig. 1.   Enzymatic ligation of synthetic oligonucleotides.

Jay et al. (1984a) in the synthesis of a human immune γ-interferon gene. A 453-base-pair DNA duplex was assembled from 66 oligonucleotides synthesized by a solid-phase phosphite approach. The oligonucleotides were purified by high-pressure liquid chromatography and sequenced individually by the technique of Maxam and Gilbert (1980). The 66 fragments were assembled into six blocks which in turn were assembled into the complete gene. Reading just this paper presents a feeling for what is possible today with these techniques.

2.   Partial Chemical Synthesis–Enzymatic Repair Approach

Escherichia coli DNA polymerase catalyzes a template-directed "repair" synthesis that serves to add specific nucleotides to a primer DNA. At 5–8°C, the incorporation reaction stops when the single-stranded section is completely repaired and no initiation of new chains occurs. Partial repair syntheses when only one, two, or three of the four deoxynucleotides triphosphates are also possible, and these are sometimes used to modify the ends of a DNA piece to be cloned. The large tryptic fragment (Klenow) of DNA polymerase is generally used to avoid $5' \rightarrow 3'$ exonucleolysis during synthesis.

Enzymatic repair of a partial duplex by *E. coli* DNA polymerase has been used in at least three cases (Scarpulla et al., 1982; Rossi et al., 1982; Buell et al., 1985) to construct synthetic genes; the strategy is to reduce the number of chemically synthesized oligonucleotides needed by making partially overlapping fragments, which are then repaired to blunt-ended complete duplexes. These blunt-ended duplexes are either cloned directly or treated with restriction endonucleases to regenerate sticky ends for cloning; linkers can also be used. The number of fragments synthesized is reduced at the expense of having to clone blunt-ended DNAs, which is more difficult than the traditional sticky-end approach.

Scarpulla et al. assembled a 63-base-pair duplex coding for the A chain of human insulin, using 8 oligonucleotides totaling 101 bases, for a gain of 20% over the conventional approach requiring synthesis of 126 bases; the blunt-ended fragment was then cloned using linkers and later retrieved with a special *Mbo*II retrieval adapter.

Rossi et al. assembled a duplex of 132 base pairs coding for part of human α_2 interferon using only four oligonucleotides totaling 163 bases, for an economy of 38% over the conventional approach requiring synthesis of 264 bases; their method made good use of the capacity of their solid-phase phosphotriester chemistry to produce oligonucleotides up to 40 bases long.

Buell et al. assembled two duplexes of 98 and 114 base pairs, respectively, coding for somatomedin C using 14 oligonucleotides totaling 333 bases, an economy of 21% over the conventional approach; each duplex was cloned separately.

Until now this method has seen comparatively little use; however, the advent of automated solid-phase DNA synthesizers able to produce fragments of 80–90 bases long might well revive interest in this technique.

B. Ligating the DNA Pieces Together

1. One-Pot Ligation

It is theoretically possible to put in one pot all the oligonucleotides corresponding to one's synthetic gene, denature by heating, reanneal, and clone the whole thing. The literature shows a few examples of researchers brave enough to follow such a route; Tanaka *et al.* (1982), who synthesized in this fashion a 44-base-pair long gene coding for α-neoendorphin, gives a typical procedure. Other examples belong to Grundstrom *et al.* (1985), Sproat and Gait (1985), and Sung *et al.* (1987).

2. Stepwise Ligation

This slower but surer approach has been used in almost all cases; again the work of Jay *et al.* (1984a) provides a good illustration. The 66 oligonucleotides were assembled in six intermediate blocks; each intermediate block was itself assembled in a stepwise fashion designed to verify each individual ligation step. Furthermore, each of the individual oligonucleotides had been sequenced by a modified Maxam and Gilbert (1980) method.

3. "In Vivtro" Ligation

Without prior *in vitro* enzymatic ligation a DNA duplex was assembled sucessfully by transforming competent cells with a mixture containing six synthetic complimentary oligodeoxynucleotides and a linearized plasmid (Narang *et al.,* 1986). The results demonstrate that enzymatic ligation systems inside the cell can be used in the joining step (see Fig. 3).

C. Cloning the Assembled DNA Piece

Once the synthetic DNA has been ligated together it becomes essentially indistinguishable from any other piece of natural DNA; in particular, the cloning and sequencing techniques so successful in molecular genetics can be applied. The reader is referred to a textbook such as Maniatis *et al.* (1982) for experimental protocols; here we will only give an outline of the methodology.

1.   Insertion in the Cloning Vehicle

Cloning a synthetic DNA involves ligating it to a larger piece of DNA capable of self-replication within a living organism (generally a phage or plasmid); the synthetic DNA is thus replicated along with the vector, giving the researcher a potentially unlimited supply of what was originally a difficult-to-produce material. The usual method is to ligate the synthetic DNA into the cohesive ends produced by digestion of the cloning vector with appropriate restriction enzymes (Fig. 2). The cohesive ends on the synthetic DNA are obtained by one of three different means; they can be synthesized as such from the beginning, they can come from a linker (Heyneker *et al.*, 1976) which has been ligated to a blunt-ended synthetic DNA and then cleaved with the appropriate restriction enzyme, or they can come from building appropriate restriction sites at the end of the blunt-ended DNA and then cleaving with the appropriate restriction enzyme site. It is, of course, also possible to ligate blunt-ended synthetic DNA in a blunt-ended cloning site on the vector; this is rarely done however, because the procedure is inefficient and the synthetic DNA is generally available only in very limited amounts.

2.   Transformation into Competent Cells and Colony Hybridization

a.   Transformation into Competent Cells. The methods used here are the same as those generally used in any recombinant DNA experiment (Maniatis *et al.*, 1982). Using synthetic DNA instead of naturally derived material in no way alters the experimental protocols.

b.   Colony Hybridization. This is both an essential and straightforward step: Essential because there are almost always colonies arising from faulty DNA pieces, arising from plasmids having inserted foreign bits of DNA, unforeseen DNA rearrangements, plasmids closing back without any inserts, and such like cloning incidents; straightforward because fortunately the synthetic approach to cloning provides the best possible probes in the form of leftover synthetic DNA fragments used in the assembly. It is good practice to screen independently with several distinct synthetic DNA pieces and to only select these colonies which hybridize to all pieces for further characterization by DNA sequencing.

3.   DNA Sequencing

Again, this is an essential step; it is not infrequent to find in the synthetic gene clones point mutations or small deletions or insertions too small to be noticeable by colony hybridization, even under stringent con-

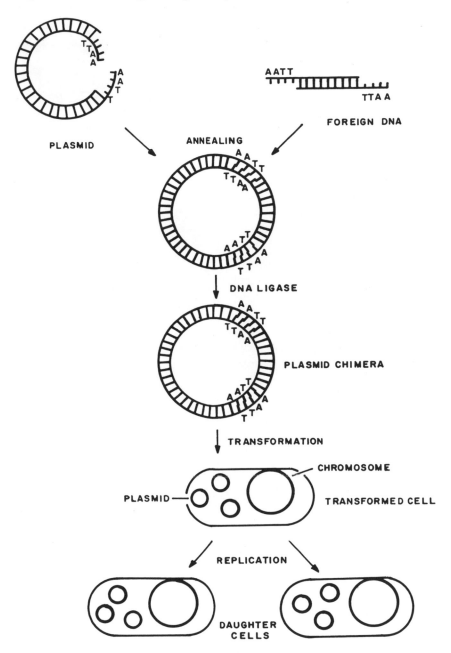

Fig. 2. Recombinant DNA methodology.

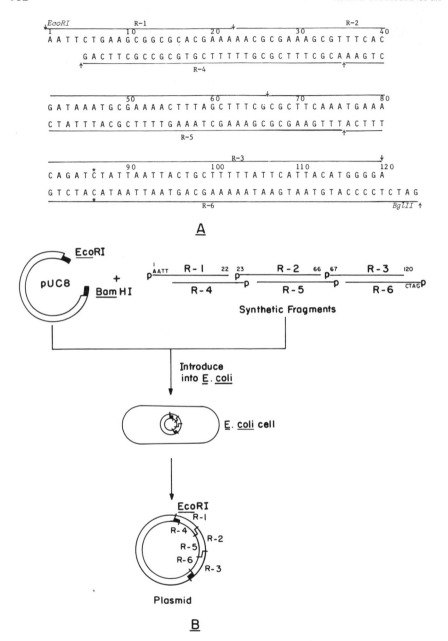

Fig. 3. "*In vivtro*" method of gene assembly.

ditions. The usual sequencing techniques of Maxam and Gilbert (1980) and Sanger *et al.* (1977) are used just like with any other DNA.

D. Hybrid Gene Synthesis

Gene synthesis as reported so far has adopted the "one synthesis, one gene" strategy. One strand of the synthetic DNA duplex would encode a polypeptide and the opposite strand would maintain the duplex structure required for subsequent biological manipulations. Recently a new approach, hybrid gene synthesis, has been developed to utilize fully both strands of the synthetic DNA duplex to encode two different polypeptides, leading to synthesis of two genes in a single operation.

1. Simultaneous Synthesis of Wild- and Mutant-Type Genes

In the synthesis of human parathyroid hormone gene and analogs, Sung *et al.* (1986a) incorporated into their design a mismatched base-pair, representing modification of a triplet codon, between the overlapping synthetic oligonucleotides. Then the oligonucleotides were phosphorylated, annealed, and ligated directly into a linearized plasmid vector, bypassing any intermediate steps of purification and gene assembly. After transformation in bacteria, the two DNA strands of the plasmid heteroduplex, as individual templates of plasmid replication, yielded two progenies bearing either the wild or mutant type of the target gene.

2. Simultaneous Synthesis of Natural Homologous Genes

With the success of the automated DNA synthesizer in preparing oligonucleotides longer than before, multiple regions of base mismatch can now be designed between the synthetic oligonucleotides without serious disruption of the essential duplex structure. These conditions permitted Sung *et al.* (1986b) to extend hybrid gene synthesis to the simultaneous assembly of natural homologous genes which differ substantially. The two homologous peptides, mouse and human epidermal growth factors (EGF), differ in 16 amino acids out of a total of 53, amounting to a difference of 30% in the amino acid sequence. For the synthetic strategy, 16 single or double base mismatches were designed in the heteroduplex intermediate, with one strand encoding the human EGF and the opposite strand indirectly encoding the mouse EGF in complementary sequence. Seven synthetic gene fragments (16- to 73-mers) were phosphorylated, annealed, and ligated directly into a linearized plasmid vector. After transformation in *E. coli* HB101 (*recA* 13), two plasmid progenies bearing either mouse or human EGF gene were obtained (Fig. 4).

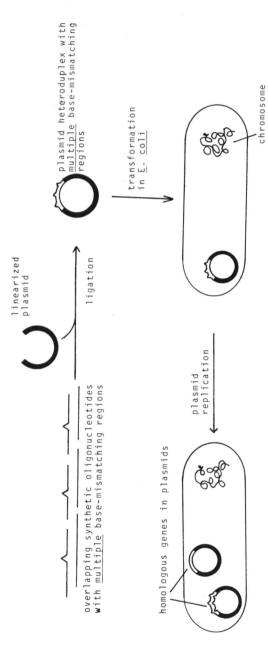

Fig. 4. Hybrid gene synthesis.

3. Synthesis of Chimera of Natural Homologous Genes

The nature of the transformation host can also determine the course of plasmid replication. When the assembly process of human and mouse EGF genes was repeated with transformation host *E. coli* JM103, a strain that favors DNA recombination, a large number of plasmid progenies had widespread crossover of mouse and human sequences in the whole EGF insert. This resulted in the formation of genuine hybrid or chimeric EGF genes. Despite the exchange of genetic messages, the reading frame of the chimeric genes was maintained in all cases examined.

It would be of great interest to see if new biologically active enzymes can be developed in form of chimera of natural homologous enzymes, via combination of favorable features of the latter. The literature of enzyme engineering shows that prediction of the effect of a specific point mutation in an enzyme is still a "hit-or-miss" affair; deliberate formation of chimeras between homologous natural or synthetic genes by the process just described would be a practical and rational way to generate a large number of chimeric enzymes for screening.

E. Expression of the Synthetic Gene

1. General Considerations on the Choice of the Expression Vector and Host

A large volume could easily be written on the various vector–host combinations used for expression of synthetic DNA. We will limit ourselves here to a brief summary of the salient features of the most frequently used *E. coli* vector systems.

The tasks performed by the expression vector in the production of a protein product are multiple, complex, and sometimes conflicting. The most important ones are as follows: (1) the vector is to maintain itself within the host population by some sort of regulated maintenance mechanism; (2) the vector must allow for a high level of transcription and translation of the foreign gene in its host organism; (3) the vector may be designed to produce the synthetic gene either free or as a fusion protein with some other, larger protein; (4) the expression vector may contain a signal peptide sequence designed to lead the protein product of the synthetic gene into the periplasmic space of the organism or outside of the cell.

We will now briefly discuss each of the control regions responsible for these functions.

2. Control Regions

a. *Maintenance of the Vector.* To maintain itself within the host cells the vector must contain an origin of replication and some sort of forced maintenance mechanism; antibiotic resistance is most often used in *E. coli*. A classic example of such a construction is the plasmid pBR322 (Bolivar *et al.*, 1977); this plasmid and its derivatives have been used very frequently by the assemblers of synthetic genes.

There are many factors which affect maintenance of a vector within its host; in particular, overexpression of a foreign gene toxic to *E. coli* can lead to loss of all or part of the vector (Brosius, 1984), or to chromosomal mutations which decrease the copy number of the plasmid (Nugent *et al.*, 1983). It has also been shown that lack of transcription termination at the end of a strongly expressed gene has a negative effect on the copy number of a plasmid (Stueber and Bujard, 1982). Thus it can be seen that there probably is no such thing as a truly universal expression vector; rather the vector must be adapted in each case to the particular gene being expressed.

b. *Vector Copy Number Regulation.* Other things being equal, production of a protein product is dependent on the number of copies of the originating gene present within the organism. Plasmids susceptible to runaway replication under induction have been used to obtain high levels of expression of foreign proteins in *E. coli*. An example of this type of work is given by Masui *et al.* (1983) working with a plasmid derived from the runaway plasmid of Uhlin *et al.* (1979); whether this approach is generally superior to the use of moderate copy number vectors remains to be determined.

c. *Transcription Initiation.* Transcription initiation in *E. coli* requires at least two conserved regions, one located approximately 10 base pairs (Pribnow, 1975; Schaller *et al.*, 1975) upstream of the transcription initiation site and one located 35 base pairs upstream of that same site (Maniatis *et al.*, 1975). Much work has been done trying to optimize both the promoter sequences (see, for instance, Amann *et al.*, 1983, among others) and the spacing between them (Russell and Bennett, 1982). Several of the currently used systems will be encountered in Table I in Section III of this chapter; the review article of Bok (1983) will also be of benefit to readers wishing to go further in the subject.

Transcription termination is separated into two types (Holmes *et al.*,

1983): factor dependent and factor independent. Factor-independent termination involves a GC-rich (often showing dyad symmetry) region followed by an AT-rich sequence and involves RNA polymerase being released from the DNA substrate; factor-dependent termination does not seem to rely on any consistent structured feature of the DNA and involves other proteins such as *rho* and *nusA*. As discussed above, lack of transcription termination can have a deleterious influence on the stability of expression vectors in *E. coli*.

d. Translation. Translation initiation in *E. coli* requires a ribosome binding site (Shine and Dalgarno, 1974) located between 3 and 12 bases upstream from the first AUG initiation codon. The distance between the Shine–Dalgarno region and the AUG has been shown (Backman and Ptashne, 1978; Shepard *et al.*, 1982) to strongly influence expression of certain genes. A synthetic Shine–Dalgarno sequence has been shown by Jay *et al.* (1982) to greatly increase expression of several genes.

Translation termination is the simplest control region to deal with in that it only requires one or more termination codons in phase with the reading frame of the synthetic gene; their presence or absence determines whether the synthetic gene will be expressed free or as a fused protein.

e. Fusion of the Desired Peptide with Another. Experience has shown that expression of small peptides into *E. coli* is generally not feasible unless the small peptide is fused to a larger one; it seems that proteases occurring naturally within *E. coli* quickly degrade small peptides. For that reason, small peptides are most often expressed as fusion proteins with a large molecule such as β-galactosidase, as seen in Table I of this chapter.

f. Fusion with Export Signal Peptides. Attempts have been made to find or create signal peptides which would direct the export of recombinant DNA-derived protein outside of *E. coli*. Chan *et al.* (1981) succeeded with preproinsulin. Ghrayeb *et al.*, 1984) have used the *OmpA* signal peptide in an attempt to create a generalized *E. coli* secretion vector; the advantages in the subsequent purification steps could be quite considerable. Until now, these efforts have not been generally successful; isolated successes have been reported, but also many failures. Each new protein seems to represent a unique case, and there are some which seem unexportable no matter what signal peptide is fused to them; an example is found in the work of Ohsuye *et al.* (1983) on α-neoendorphin.

TABLE I

Reported Gene Syntheses

Type of gene, authors, and year of publication	Length (bp)	Summary of synthesis and assembly	Host; vector; fusion protein; promoter; yield of protein
A. Small genes (30–90 bases long)			
Bradykinin, Korobko et al. (1979)	41	Solution phosphotriester chemistry; 8 fragments sequenced by two-dimensional method of Jay et al. (1974)	E. coli; pBR322 derivative with long β-galactosidase fragment; lac promoter; 1 mg of bradykinin per 25 g of cells by radio immunoassay (RIA) and bioassay; see Gorodetskii et al. (1980)
α-Neoendorphin, Tanaka et al. (1982)	44	Solution phosphotriester chemistry; 8 fragments sequenced by two-dimensional method; one-step assembly	E. coli; pBR322 derivative with long β-galactosidase fragment; lac promoter; 4 mg/10.9 g wet cells; E. coli; pBR322 derivative with alkaline phosphatase promoter and 94% of alkaline phosphatase gene; 60% of total cellular protein; no export to periplasm; Ohsuye et al. (1983)
Angiotensin I, Kumarev et al. (1980)	48	Solution triester; 6 fragments; one-step assembly, with subsequent linker addition	E. coli; pBR322 derivative with long β-galactosidase fragment; lac promoter; 300 pg/10^8 cells; E. coli; λplac5 phage; 3 ng/10^8 cells
Somatostatin, Itakura et al. (1977)	56	Solution triester; 8 fragments; stepwise assembly	E. coli; pBR322 derivative with long β-galactosidase fragment; lac promoter; 0.03% of total cellular protein by RIA
B. Medium-sized genes (90–200 base pairs)			
$α_1$-Thymosin, Wetzel et al. (1980)	98	Solution triester; 16 fragments; stepwise assembly	E. coli; pBR322 derivative with long β-galactosidase fragment; lac UV5 promoter; 200 mg/kg of cells
Human insulin A and B chains, Crea et al. (1978)	77 and 104	Solution triester; 29 fragments; stepwise, separate assembly of A and B chains	E. coli; pBR322 derivative with long β-galactosidase fragment; lac promoter; 10 mg/24 g wet weight of cells
Secretin, Miyoshi et al. (1981)	110	Solid-phase triester; 16 fragments; stepwise (3 blocks)	E. coli; pBR322; 182 amino acid β-lactamase fragment; bla promoter; 7 ng from a 3-liter culture; see Suzuki et al. (1982)

TABLE I (*Continued*)

Type of gene, authors, and year of publication	Length (bp)	Summary of synthesis and assembly	Host; vector; fusion protein; promoter; yield of protein
β-Urogastrone Urdea *et al.* (1983)	170	Solid-phase phosphoramidite; 10 fragments; stepwise in 3 blocks	*S. cerevisiae;* no fusion fragment; yeast glyceraldehyde-3-phosphate dehydrogenase (GAPDH) promoter; 30 μg/liter of culture
Smith *et al.* (1982)	174	Solution triester; 23 fragments; stepwise in two blocks	*E. coli*; pAT153 derivative; 7-amino-acid *trpE* fragment; *trp* promoter; 2–3 mg/liter of culture
"Mini-C" analog of human proinsulin, Wetzel *et al.* (1981)	185	Solution triester; 32 fragments; stepwise	*E. coli*; pBR322 derivative with long β-galactosidase fragment; *lac* promoter; 12 mg from 70 liters of culture medium

C. Large sized genes (200 and more base pairs)

Somatomedin C, Buell *et al.* (1985)	212	Solid-phase triester; 14 fragments; stepwise in two blocks	*E. coli*; pUC8; short β-galactosidase fragments and/or synthetic sequences; P_L and *lac* promoters; 65 ng/10^7 cells
Human complement fragment C5a, Mandecki *et al.* (1985)	253	Solid-phase phosphoramidite; 16 fragments; stepwise assembly in three subfragments	*E. coli*; pBR322 derivative; no fusion peptide; *lac* UV5-D promoter; 0.007% of total bacterial protein
Human-like proinsulin and preproinsulin, Brousseau *et al.* (1982)	277, 355	Solution triester; 41 and 53 fragments; stepwise assembly	*E. coli*; pBR322 derivative; 590-amino-acid fragment of β-galactosidase; *lac* promoter; 41.7 μg/mg of total protein by RIA (Guo *et al.*, 1984) *S. cerevisiae;* 280-amino-acid yeast galactokinase; *GAL1* promoter; 119 ng/mg of total protein by RIA (Stepien *et al.*, 1982); see also Shen (1984) and Georges *et al.* (1984) for further expression results
Human proinsulin, Ovchinnikov *et al.* (1984)	286	Solution triester; 42 fragments	No data reported

(*continued*)

TABLE I (*Continued*)

Type of gene, authors, and year of publication	Length (bp)	Summary of synthesis and assembly	Host; vector; fusion protein; promoter; yield of protein
Ribonuclease S, Nambiar *et al.* (1984)	330	Solid-phase phosphoramidite; 66 fragments; stepwise in 8 blocks	No data reported
Human γ-interferon Engels *et al.* (1984)	452	Solid-phase phosphoramidite; 34 fragments; stepwise assembly; use of HPLC in purification	*E. coli*; pUC8 derivative; no fusion fragment; *tac* promoter; 1 mg/liter per OD$_{578}$
Jay *et al.* (1984a,b)	453	Solid-phase phosphoramidite; 66 fragments; stepwise assembly; use of HPLC in purification	*E. coli*; pBR322 derivatives; no fusion fragment; *bla* and T5 promoters; synthetic ribosomal binding sites; 4×10^9 units/ liter of culture;
Tanaka *et al.* (1982)	454	Solid-phase triester; 62 fragments; stepwise assembly; use of HPLC for purification	*E. coli*; pBR322 derivative; no fusion fragment; *lac* UV5 and alkaline phosphatase promoters; 15% of total protein (alkaline phosphatase promoter)
Human β-interferon, Nagase *et al.* (1983)	510	Solid-phase triester; 61 fragments	*E. coli*; *trp* promoter; no numbers given on expression level
Human α$_2$-interferon, Edge *et al.* (1983)	511	Solid-phase triester; 68 fragments; use of HPLC; stepwise in 13 parts	*E. coli*; pAT153 derivative; *lac* UV5 and *trp* promoters; 5×10^7 units/liter of culture at 1 OD unit
Human α$_1$-interferon, Edge *et al.* (1981)	514	Solid-phase triester; 67 fragments; HPLC purification; stepwise assembly in 11 blocks	*E. coli* and *Methylophilus methylotrophus; lac* UV5 promoter; 19,000–38,000 units/0.15 ml of bacterial extract; see DeMaeye *et al.* (1982)
Human growth hormone, Ikehara *et al.* (1984)	584	Solid-phase triester; 78 fragments; HPLC purification; stepwise assembly in 8 blocks	*E. coli*; pBR322 derivative; no fusion fragment; *trp* promoter; 169 μg/ml of culture medium
Poly(L-aspartyl-L-phenyl-alanine), Doel *et al.* (1980)	Up to 900	Polymerization of a small basic unit of 12 bases made by solution triester	*E. coli*; pBR322 derivative; *trp* promoter; 10^4–10^5 molecules per cell

III. REPORTED GENE SYNTHESES

We have chosen to present in Table I the reported syntheses in order of increasing size; this allows us to demonstrate better the different construction features used, whether the gene is small and needs to be fused to a larger protein to stabilize it or whether expression is sought of the native peptide without any fused sequences. The division into small (less than 60 base pairs), medium (between 90 and 200 base pairs), and large (more than 200 base pairs) corresponds in our minds to the different strategies used for expression; peptides below 20 amino acids are invariably expressed as fused proteins, those between 20 and 70 amino acids are most often expressed as fused proteins but can also be expressed independently, while those over 70 amino acids are generally expressed independently.

The expression studies mentioned here are those found in the original publications describing gene synthesis or in publications referred to in these original publications. We did not attempt to single out, from the large amount of work being done on advanced expression vectors, those publications where a vector was incidentally used for a synthetic gene.

IV. CONCLUSION

Twenty-five years ago, although the technology of DNA synthesis was still in its infancy, it made a crucial contribution to the greatest scientific achievement of that time, the elucidation of the genetic code. Now that DNA synthesis has grown in scope from repeating triplet sequences to complete genes of 600 bases or more (and there are reports of genes 1800 bases or more being currently synthesized), the challenge ahead lies in elucidating that second-order genetic code by which the linear sequence of bases encodes a given activity or structure within the organism. That means we must elucidate how the primary structure of a protein determines its secondary, tertiary, and quaternary structure, its location within the cell, and its activity. The technology of synthetic genes, coupled with its sister technology site-specific mutagenesis, will have great application since it will allow research scientists to create numerous variants of existing genes modified in precisely defined ways. This technology also introduces the possibility of designing new, heretofore unknown genes to extend further the range of tasks that microorganisms can undertake on our behalf.

REFERENCES

Agarwal, K. L., Buchi, H., Caruthers, M. H., Gupta, N., Khorana, H. G., Kleppe, K., Kumar, A., Ohtsuka, E., RajBhandary, U. L., van de Sande, J. H., Sgaramella, V., Weber, H., and Yamada, T. (1970). *Nature (London)* **227,** 27–34.

Amann, E., Brosius, J., and Ptashne, M. (1983). *Gene* **25,** 167–178.

Backman, K., and Ptashne, M. (1978). *Cell (Cambridge, Mass.)* **13,** 65–71.

Bok, S. H. (1983). *Dev. Ind. Microbiol.* **24,** 255–270.

Bolivar, F., Rodriguez, R. L., Greene, P. J., Betlach, M. C., Heyneker, H. L., and Boyer, H. W. (1977). *Gene* **2,** 95–113.

Brosius, J. (1984). *Gene* **27,** 161–172.

Brousseau, R., Scarpulla, R. C., Sung, W., Hsiung, H., Narang, S. A., and Wu, R. (1982). *Gene* **17,** 279–289.

Buell, G., Schulz, M. F., Selzer, G., Chollet, A., Movva, N. R., Semon, D., Escanez, S., and Kawashima, E. (1985). *Nucleic Acids Res.* **13,** 1923–1938.

Chan, S. J., Weiss, J., Konrad, M., White, T., Bahl, C., Yu, S. D., Marks, D., and Steiner, D. F. (1981). *Proc. Natl. Acad. Sci. U.S.A.* **78,** 5401–5405.

Crea, R., Kraszewski, A., Hirose, T., and Itakura, K. (1978). *Proc. Natl. Acad. Sci. U.S.A.* **75,** 5765–5769.

DeMaeyer, E., Skup, D., Prasad, K. S. N., DeMaeyer-Guignard, J., Williams, B., Meacock, P., Sharpe, G., Pioli, D., Hennam, J., Schuch, W., and Atherton, K. (1982). *Proc. Natl. Acad. Sci. U.S.A.* **79,** 4256–4259.

Doel, M. T., Eaton, M., Cook, E. A., Lewis, H., Patel, T., and Carey, N. H. (1980). *Nucleic Acids Res.* **8,** 4575–4591.

Edge, M. D., Greene, A. R., Heathcliffe, G. R., Meacock, P. A., Schuch, W., Scanlon, D. B., Atkinson, T. C., Newton, C. R., and Markham, A. F. (1981). *Nature (London)* **292,** 756–762.

Edge, M. D., Greene, A. R., Heathcliffe, G. R., Moore, V. E., Faulkner, N. J., Camble, R., Petter, N. N., Trueman, P., Schuch, W., Henman, J., Atkinson, T. C., Newton, C. R., and Markham, A. F. (1983). *Nucleic Acids Res.* **11,** 6419–6435.

Engels, J., Leineweber, M., and Uhlmann, E. (1984). *Angew. Chem.* **96,** 969–970.

Georges, F., Brousseau, R., Michniewicz, J., Prefontaine, G., Stawinski, J., Sung, W., Wu, R., and Narang, S. A. (1984). *Gene* **27,** 201–211.

Ghrayeb, J., Kimura, H., Takahara, M., Hsiung, H., Masui, Y., and Inouye, M. (1984). *EMBO J.* **3,** 2437–2442.

Goeddel, D. V., Heyneker, H. L., Hozumi, T., Arentzen, R., Itakura, K., Yansura, D. G., Ross, M. J., Miozzari, G., Crea, R., and Seeburg, P. H. (1979b). *Nature (London)* **281,** 544–548.

Gorodetskii, S. I., Kapelinskaya, T. V., Lisenkov, A. F., Slyusarenko, A. G., and Dubinin, N. P. (1980). *Dokl. Akad. Nauk. SSR* **250**(1), 208–212.

Grundstörm, T., Zenke, W. M., Wintzerith, M., Matthes, H. W. D., Staub, A., and Chambon, P. (1985). *Nucleic Acids Res.* **13,** 3305–3316.

Guo, L. H., Stepien, P. P., Tso, J. Y., Brousseau, R., Narang, S. A., Thomas, D. Y., and Wu, R. (1984). *Gene* **29,** 251–254.

Heyneker, H. L., Shine, J., Goodman, H. M., Boyer, H. W., Rosenberg, J., Dickerson, R. E., Narang, S. A., Itakura, K., Lin, S., and Riggs, A. D. (1976). *Nature (London)* **263,** 748–752.

Holmes, W. M., Platt, T., and Rosenberg, M. (1983). *Cell (Cambridge, Mass.)* **32,** 1029–1032.

Ikehara, M., Ohtsuka, E., Tokunaga, T., Taniyama, Y., Iwai, S., Kitano, K., Miyamoto, S., Ohgi, T., Sakuragawa, Y., Fujiyama, K., Ikari, T., Kobayashi, M., Miyake, T., Shibahara, S., Ono, A., Ueda, T., Tanaka, T., Baba, H., Miki, T., Sakurai, A., Oishi, T., Chisaka, O., Matsubara, K. (1984). *Proc. Natl. Acad. Sci. U.S.A.* **81**, 5956–5960.
Itakura, K., Hirose, T., Crea, R., Riggs, A. D., Heyneker, H. L., Bolivar, F., and Boyer, H. W. (1977). *Science* **198**, 1056–1063.
Jay, E., Bambara, A., Padmanabhan, R., and Wu, R. (1974). *Nucleic Acids Res.* **1**, 331–342.
Jay, E., Seth, A. K., Rommens, J., Sood, A., and Jay, G. (1982). *Nucleic Acids Res.* **10**, 6319–6329.
Jay, E., MacKnight, D., Lutze-Wallace, C., Harrison, D., Wishart, P., Liu, W. Y., Asundi, V., Pomeroy-Cloney, L., Rommenes, J., Eglington, L., Pawlak, J., and Jay, F. (1984a). *J. Biol. Chem.* **259**, 6311–6317.
Jay, E., Rommens, J., Pomeroy-Cloney, L., MacKnight, D., Lutze-Wallace, C., Wishart, P., Harrison, D., Liu, W. Y., Asundi, V., Dawood, M., and Jay, F. (1984b). *Proc. Natl. Acad. Sci. U.S.A.* **81**, 2290–2294.
Khorana, H. K., Agarwal, K. L., Besmer, P., Buchi, H., Caruthers, M. H., Cashion, P. J., Fridkin, M., Jay, E., Kleppe, K., Kleppe, R., Kumar, A., Loewen, P. C., Miller, R. C., Minamoto, K., Panet, A., RajBhandary, U. L., Ramamoorthy, B., Sekiya, T., Takeya, T., and van de Sande, J. H. (1976). *J. Biol. Chem.* **251**, 565–570.
Korobko, V. G., Dobrynin, V. N., Boldyreva, E. F., Severtsova, I. V., Chernov, B. K., Kolosov, M. N., Gorodetskii, S. I., Slyusarenko, A. G., Kapelinskaya, T. V., Lisenkov, A. F., and Dubinin, N. P. (1979). *Bioorg. Khim.* **5**, 1802–1815.
Kumarev, V. P., Rivkin, M. I., Bogachev, V. S., Baranova, L. V., Merkulov, V. M., Rybakov, V. N., Solenov, E. I., and Fedorov, V. I. (1980). *Dokl. Akad. Nauk SSSR* **252**, 1506–1510.
Mandecki, W., Mollison, K. W., Bolling, T. J., Powell, B. S., Carter, G. W., and Fox, J. L. (1985). *Proc. Natl. Acad. Sci. U.S.A.* **82**, 3543–3547.
Maniatis, T., Ptashne, M., Backman, K., Kleind, D., Flashman, S., Jeffrey, A., and Maurer, R. (1975). *Cell (Cambridge, Mass.)* **5**, 109–113.
Maniatis, T., Fritsch, E. F., and Sambrook, J. (1982). "Molecular Cloning: A Laboratory Manual." Cold Spring Harbor Lab., Cold Spring Harbor, New York.
Marians, K. J., Wu, R., Stawinski, J., Hozumi, T., and Narang, S. A. (1976). *Nature (London)* **263**, 744–748.
Masui, Y., Coleman, J., and Inouye, M. (1983). *in* "Experimental Manipulation of Gene Expression" (M. Inouye, ed.), pp. 15–32. Academic Press, New York.
Maxam, A. M., and Gilbert, W. (1980). *in* "Methods in Enzymology" (L. Grossman and K. Moldave, eds.), Vol. 65, pp. 499–560. Academic Press, New York.
Miyoshi, K., Hasegawa, A., Tomiyama, M., and Miyake, T. (1981). *Nucleic Acids Symp. Ser.* **10**, 197–200.
Nagase, Y., Nakamura, N., Tohyama, J., Watanabe, S., Ogino, H., Horikoshi, K., Nii, A., Soma, N., Nobuhara, M., Suzuki, Y., and Mochida, E. (1983). *Nucleic Acids Symp. Ser.* **12**, 83–86.
Nambiar, K. P., Stackhouse, J., Stauffer, D. M., Kennedy, W. P., Eldredge, J. K., and Benner, S. A. (1984). *Science* **223**, 1299–1300.
Narang, S. A., Dubuc, G., Yao, F. L., and Michniewicz, J. J. (1986). *Biochem. Biophys. Res. Commun.* **134**, 407–411.
Nugent, M. E., Primrose, S. B., and Tacon, W. C. A. (1983). *Dev. Ind. Microbiol.* **24**, 271–285.
Ohsuye, K., Nomura, M., Tanaka, S., Kubota, I., Nakazato, H., Shinagawa, H., Nakata, A., and Noguchi, T. (1983). *Nucleic Acids Res.* **11**, 1283–1294.

Ovchinnikov, Yu. A., Efimov, V. A., Ivanova, I. N., Reverdatto, S. V., Skiba, N. P., and Chakhmakhcheva, O. G. (1984). *Gene* **31,** 65–78.

Pribnow, D. (1975). *Proc. Natl. Acad. Sci. U.S.A.* **79,** 1069–1072.

Rossi, J. J., Kierzek, R., Huang, T., Walker, P. A., and Itakura, K. (1982). *J. Biol. Chem.* **257,** 9226–9229.

Russell, D. R., and Bennett, G. N. (1982). *Gene* **20,** 231–243.

Sanger, F., Nicklen, S., and Coulson, A. R. (1977). *Proc. Natl. Acad. Sci. U.S.A.* **74,** 5463–5467.

Scarpulla, R. C., Narang, S. A., and Wu, R. (1982). *Anal. Biochem.* **121,** 356–365.

Schaller, H., Gray, C., and Herrmann, K. (1975). *Proc. Natl. Acad. Sci. U.S.A.* **72,** 737–741.

Shen, S. H. (1984). *Proc. Natl. Acad. Sci. U.S.A.* **81,** 4627–4631.

Shepard, H. M., Yelverton, E., and Goeddel, D. V. (1982). *DNA* **1,** 125–131.

Shine, J., and Dalgarno, L. (1974). *Proc. Natl. Acad. Sci. U.S.A.* **71,** 1342–1346.

Smith, J., Cook, E., Fotheringham, I., Pheby, S., Derbyshire, R., Eaton, M. A. W., Doel, M., Lilley, D. M. J., Pardon, J. F., Patel, T., Lewis, H., and Bell, L. D. (1982). *Nucleic Acids Res.* **10,** 4467–4482.

Sproat, B. S., and Gait, M. J. (1985). *Nucleic Acids Res.* **13,** 2959–2977.

Stepien, P. P., Brousseau, R., Wu, R., Narang, S. A., and Thomas, D. Y. (1983). *Gene* **24,** 289–297.

Stueber, D., and Bujard, H. (1982). *EMBO J.* **11,** 1399–1404.

Sung, W. L., Zahab, D. M., Yao, F. L., and Tam, C. S. (1986a). *Biochem. Cell Biol.* **64,** 133–138.

Sung, W. L., Zahab, D. M., Yao, F. L., Wu, R., and Narang, S. A. (1986b). *Nucleic Acid Res.* **14,** 6159–6168.

Suzuki, M., Sumi, S., Hasegawa, A., Nishizawa, T., Miyoshi, K., Wakisawa, S., Miyake, T., and Misoka, F. (1982). *Proc. Natl. Acad. Sci. U.S.A.* **79,** 2475–2479.

Tanaka, S., Oshima, O., Ohsue, K., Ono, T., Oikawa, S., Takano, I., Noguchi, T., Kangawa, K., Minamino, N., and Matsuo, H. (1982). *Nucleic Acids Res.* **10,** 1741–1754.

Uhlin, B. E., Molin, S., Gustafsson, P., and Nordstrom K. (1979). *Gene* **6,** 91–106.

Urdea, M. S., Merryweather, J. P., Mullenbach, G. T., Coit, D., Heberlein, U., Valenzuela, P., and Barr, P. J. (1983). *Proc. Natl. Acad. Sci. U.S.A.* **80,** 7461–7465.

Wetzel, R., Heyneker, H. L., Goeddel, D. V., Jhurani, P., Shapiro, J., Crea, R., Low, T. L. K., McClure, J. E., Thurman, G. B., and Goldstein, A. L. (1980). *Biochemistry* **19,** 6096–6104.

Wetzel, R., Kleid, D. G., Crea, R., Heyneker, H. L., Yansura, D. G., Hirose, T., Kraszewski, A., Riggs, A. D., Itakura, K., and Goeddel, D. V. (1981). *Gene* **16,** 63–71.

5

Chemical Synthesis of RNA

EIKO OHTSUKA
SHIGENORI IWAI
Faculty of Pharmaceutical Sciences
Hokkaido University
Sapporo 060, Japan

I. INTRODUCTION

The synthetic methodology for the synthesis of short oligoribonucleotides by the phosphodiester approach was developed in the 1960s
(Khorana, 1968) and sixty-four ribotriplets were synthesized as minimum
units of messenger RNA (mRNA) (Lohrmann *et al.*, 1966). Organic chemical syntheses of larger molecules of oligoribonucleotides have been attempted by using the phosphodiester, phosphotriester, or phosphite trics-
ter methods as reviewed in articles by Reese (1978), Ikehara *et al.* (1979),
Ohtsuka *et al.* (1982b), and Narang (1983). However, the discovery of
RNA ligase (Silver *et al.*, 1972; Gumport and Uhlenbeck, 1981) has extended the possibilities for synthesizing RNA molecules such as tRNA
(Ohtsuka *et al.*, 1981a; Wang, 1984). In the synthesis of RNA it is necessary to protect the 2'-hydroxyl group of the ribonucleoside and this is
considered more difficult than in DNA synthesis. The solid-phase synthesis of oligodeoxyribonucleotides has been facilitated by improvements in
the methodologies used. The number of reports on the synthesis of oligoribonucleotides has been much smaller than those on the deoxyribo
series.

In this chapter chemical syntheses of oligoribonucleotides will be reviewed focusing on aspects of biochemical interests.

Synthesis and Applications
of DNA and RNA

II. PROTECTING GROUPS AND CONDENSING REAGENTS FOR CHEMICAL SYNTHESIS OF OLIGORIBONUCLEOTIDES

A. Protecting Groups for Bases

The amino group of cytidine has been known to be a stronger nucleophile than sugar hydroxyl groups in condensing reactions (Gilham and Khorana, 1958). Other heterocyclic amino groups in adenine and guanine are not as reactive as sugar hydroxyl groups involved in condensing reactions. However, it is generally advisable to protect the amino groups which are not intended for phosphorylation. Protecting groups usually increase the solubility of nucleotide intermediates in organic solvents and this assists handling during purification procedures. A variety of protecting groups have been examined since the 1950s; however, the acyl groups used for protecting the amino function in the phosphodiester approach (Khorana, 1968) have proved most useful. In the deoxyribo series, various protecting groups have been examined to find more selective conditions for avoiding degradation of the glycosidic linkage of purine deoxyribosides. In the ribo series, the amino group of adenosine can be protected by treatment with either benzoyl chloride or benzoic anhydride followed by selective removal of the O-benzoyl groups and an extra benzoyl group on the base (Rammler and Khorana, 1962; Lohrmann and Khorana, 1964; Hall, 1964). These benzoylated adenosine derivatives are identified by measurement of ultraviolet absorption. The amino group of cytidine can be selectively acylated by treatment with acid anhydride in the absence of basic catalysts (Watanabe and Fox, 1966; Sasaki and Mizuno, 1967). The anisoyl group was used to maintain the stability during syntheses and final removal from oligonucleotides (Ralph and Khorana, 1961). However, N-benzoylcytidine was found to be stable under conditions for selective O-deacylation with alkali, which gave a purer product than direct N-acylation. To protect the amino group of guanosine the easily removable acetyl group was used in the 1960s. N-Isobutyrylguanosine was prepared (Ohtsuka et al., 1978a) by a procedure similar to that used for deoxyguanosine (Weber and Khorana, 1972). Schemes for preparation of N-acyl nucleosides are shown in Fig. 1. One-pot syntheses for N-benzoyldeoxyadenosine, N-benzoyldeoxycytidine (Ti et al., 1982), and N-isobutyryldeoxyguanosine (McGee et al., 1983) can potentially be used for ribonucleosides. The use of other acyl groups (Köster et al., 1981) and benzyloxycarbonyl (Watkins et al., 1982) for protecting the heterocyclic amino functions have been described.

Acid-labile trityl derivatives (Smith et al., 1962) have also been used for the exocyclic amino groups (Shimidzu and Letsinger, 1968). Dimethyl-

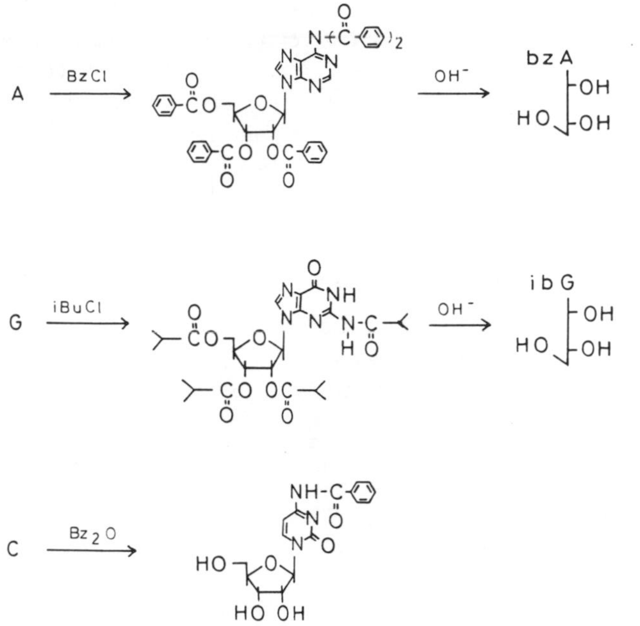

Fig. 1. Protection of the heterocyclic amino groups in the preparation of N-acyl nucleosides.

aminomethylene protection of the amino group can be removed easily under acid or alkaline conditions (Zemlicka and Holy, 1967). This type of protection proved to be too unstable for oligonucleotide syntheses. For the preparation of 5'-O-dimethoxytritylguanosine, the dimethylamino-methylene group was used for temporary protection.

The keto group of guanine or uracil was found to be modified when excess phosphodiesters were activated by arenesulfonyl triazolides or tetrazolides (Reese and Ubasawa, 1980a,b; Jones *et al.*, 1980) as shown in Fig. 2. Protection of the keto group of thymine and guanine with phenyl

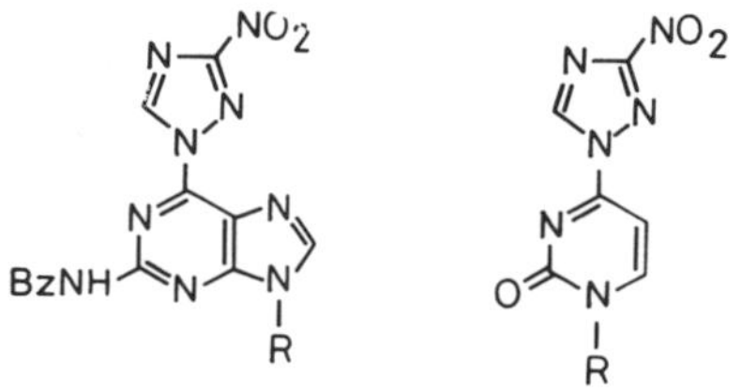

Fig. 2. Modification of the keto group with arenesulfonyl triazolides.

derivatives has been investigated (Jones *et al.*, 1981; Reese and Skone, 1984). Diphenylcarbamoyl derivatives (Kamimura *et al.*, 1984) and *p*-nitrophenyl (Himmelsbach *et al.*, 1984) and 3,4-dimethoxybenzyl (Takaku *et al.*, 1984) groups were also reported as protecting groups for the 4-O or 6-O functions of uracil, thymine, or guanine. Introduction of the acyl groups into the N-3 position of uracil has been investigated (Welch and Chattopadhyaya, 1983; Matsuzaki *et al.*, 1984).

B. Protecting Groups for Secondary Hydroxyl Functions

In the synthesis of oligoribonucleotides, the protection of the 2'-hydroxyl group in combination with that of the 5'-hydroxyl function is the essential problem. the 2'-protecting groups are usually kept until the last step in deprotection.

1. Acyl Groups

As a terminal unit the 2'- and 3'-hydroxyl groups of nucleoside can be acetylated or benzoylated via 5'-protected intermediates (Lohrmann and Khorana, 1964). In the presence of the 3'-phosphate, 2'-O-acylation of ribonucleotides was performed in the phosphodiester synthesis (Lohrmann *et al.*, 1966). Monoacylation of the secondary hydroxyl function of the nucleoside has been investigated using an orthoester. However, migration of acyl groups has been shown (Reese and Trentham, 1965; Griffin *et al.*, 1966) in aqueous pyridine and some crystallizable 3'-acyl nucleosides have been obtained as a consequence at equilibrium (Fromageot *et al.*, 1967). 2'-O-Benzoylated nucleosides were used in the phosphoramidite method and migration of the 3' → 5' internucleotide linkage to the 2' → 5' linkage was observed after treatment with ammonia when deprotection of the methyl phosphotriester was still incomplete (Kempe *et al.*, 1982).

2. Acetals and Ketals

Acid-labile protecting groups for the bicinal hydroxy function have been widely used. Ethoxymethylidene and related groups have been introduced as easily removable groups rather than the classical isopropylidene group (Chladek *et al.*, 1966; Griffin *et al.*, 1967).

Acid-labile ketals have been investigated extensively as a protecting group for the 2'-hydroxy function. In the phosphodiester approach, the tetrahydropyranyl group was introduced into 3'-phosphorylated nucleosides (Smith *et al.*, 1962) and the 2'-tetrahydropyranyl groups of the oligonucleotides were removed with acid, leaving a trace of 2' → 5' internucleotidic linkages in the products. However, contamination of the

2'-phosphorylated starting nucleosides was not excluded as a possibility in those experiments. Later treatment of oligoribonucleotides with 0.01 M hydrochloric acid at pH 2 was found not to affect internucleotidic linkages (Norman *et al.*, 1984). More acid-labile, achiral 4-methoxytetrahydro-pyran-4-yl (methoxytetrahydropyranyl, MTHP) has been introduced (Reese *et al.*, 1967) and used for oligoribonucleotide syntheses (Reese, 1978, 1985). Preparation of the reagent, 4-methoxy-3,4-dihydropyran, requires several reaction steps. 2'-*O*-Tetrahydrofuranyl nucleosides which have an acid lability similar to methoxytetrahydropyranyl nucleosides can be prepared with readily available 2,3-dihydrofuran (Ohtsuka *et al.*, 1983a,b) and used in the synthesis of an oligoribonucleotide with a chain length of 33 (Ohtsuka *et al.*, 1984). 2'-*O*-Protecting groups are prepared via 3',5'-protected nucleosides obtained by using dichloro-1,1,3,3-te-traisopropyldisiloxane (Markiewicz, 1979; Markiewicz *et al.*, 1980) as shown in Fig. 3.

3. Benzyl Derivatives

Benzyl ether has been examined as a stable protecting group for the 2'-hydroxy function. Hydrogenolysis of the 2'-*O*-benzyl nucleotides was found to be an unsatisfactory procedure probably due to the poor recovery of nucleotides from the surface of the catalyst or because of the addition of hydrogen to the 5,6-double bond of pyrimidine bases. 2'-*O*-(*o*-Nitrobenzyl) nucleosides have been prepared using *o*-nitrobenzyl bromide (Ohtsuka *et al.*, 1974, 1977a,b) or *o*-nitrophenyl diazomethane (Bartholomew and Broom, 1975; Ohtsuka *et al.*, 1981b) and have been used for the synthesis of oligonucleotides with chain lengths up to 20 (Ohtsuka *et al.*, 1979, 1981c,d). However, removal of the *o*-nitrobenzyl ether by irradiation with ultraviolet light of wavelengths longer than 280 nm was found to be about 95% complete. It is difficult to deblock larger molecules completely.

Another substituted benzyl group, *p*-methoxybenzyl, has been introduced to protect the 2'-hydroxyl function (Takaku and Kamaike, 1982)

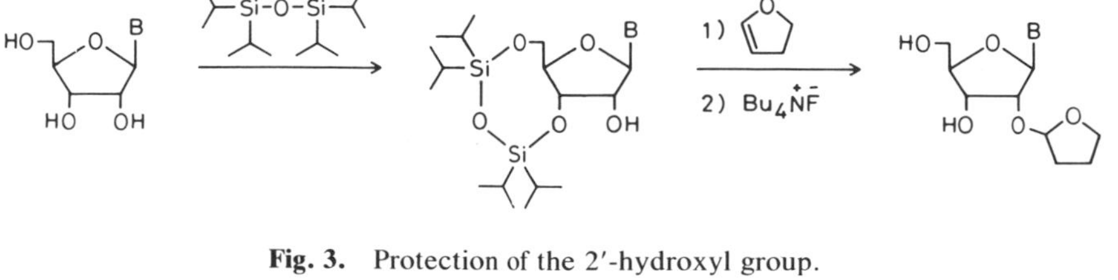

Fig. 3. Protection of the 2'-hydroxyl group.

and removal with dichlorodicyanoquinone (Oikawa *et al.*, 1982) was also examined.

4. Silyl Derivatives

The *tert*-butyldimethylsilyl group, which was originally used in the synthesis of prostaglandins, has been introduced as a protecting group for the 2'-hydroxyl function (Ogilvie, 1983). Alkaline-catalyzed migration of 2'-*O*-(*tert*-butyldimmethylsilyl) ether has been detected during the isolation process (Jones and Reese, 1979). However, rapid phosphorylation of a vicinal hydroxyl function could prevent actual contamination with the side product in the phosphite method (Ogilvie and Theriault, 1979). The 2'-*O*-(*tert*-butyldimethylsilyl) nucleosides have also been used in the phosphotriester synthesis. The stability of internucleotide linkages during treatment with tetra-*n*-butylammonium fluoride for the removal of the protecting group is not completely known.

C. Protecting Groups for the Primary Hydroxyl Group

1. Acid-Labile Groups

Acid-labile triphenylmethyl (trityl) and its *p*-methoxy derivatives (Smith *et al.*, 1962) have been used most frequently in the synthesis of oligonucleotides. Monomethoxytrityl group was found to be removed with 80% acetic acid at room temperature and suitable for selective removal from N-protected ribonucleotides, in contrast to protected deoxynucleotides which require milder acidic conditions to prevent cleavage of *N*-benzoyladenine glycosides. In the presence of acid-labile 2'-*O*-protecting groups, zinc bromide could be used for dedimethoxytritylation (Ohtsuka *et al.*, 1983b) under conditions employed in the oligodeoxyribonucleotide synthesis (Matteucci and Caruthers, 1980; Kierzek *et al.*, 1981). 9-Phenylxanthen-9-yl group was introduced as a 5'-protecting group and showed acid lability similar to that of the dimethoxytrityl protecting group (Chattopadhyaya and Reese, 1978).

2. Alkaline-Labile Groups

Acetyl and benzoyl groups can be used as a terminal blocking group. For selective removal of 5'-protecting group, more alkaline-labile acyl groups such as methoxy- and phenoxyacetyl were examined (Reese *et al.*, 1975). A bulky acyl group, trityloxyacetyl (trac) could selectively be introduced to the primary hydroxyl function and used in combination with acid-labile 2'-*O*-tetrahydropyranyl nucleosides (Werstiuk and Neilson, 1972). The rate of general base-catalyzed hydrolysis with dilute ammonia of the 5'-*O*-trac seemed to be affected by the structure of the nucleoside, and the rate decreased with the size of oligonucleotides (Neilson *et al.*, 1975).

3. Other Protecting Groups

5'-*O*-Levulinyl nucleosides have been prepared for oligonucleotide synthesis and the levulinyl group was removed by a mild treatment with hydrazine without damaging *N*-benzoylcytidine or adenosine (van Boom and Burgers, 1975, 1978). Hydrazinolysis or β-benzoylpropionyl nucleosides which had been reported earlier (Letsinger *et al.*, 1967) required stronger conditions and N-debenzoylation was detected during treatment with hydrazine in pyridine–acetic acid.

The *o*-dibromomethylbenzoyl group has been introduced as a 5'-*O*-protecting group and removed with silver perchlorate (Chattopadhyaya *et al.*, 1979). The rate of removal and the instability with phenyl sulfoxide have been discussed together with selective removal of the 5'-*O*-2-(methylthiomethoxymethyl) benzoyl group with mercury perchlorate (Reese, 1985). Some protecting groups for hydroxyl functions are shown in Fig. 4.

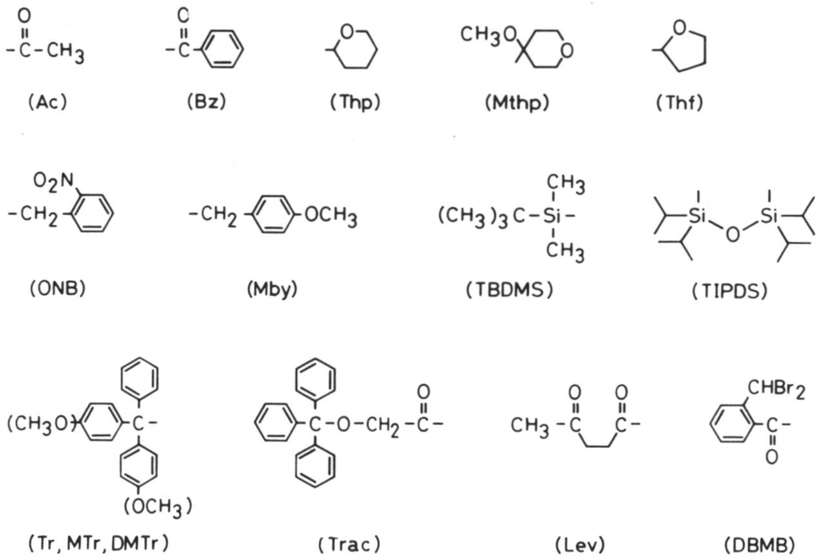

Fig. 4. Protecting groups for sugar hydroxyl functions. Abbreviations: Ac, acetyl; Bz, benzoyl; Thp, tetrahydropyran-2-yl; Mthp, 4-methoxytetrahydropyran-4-yl; Thf, tetrahydrofuran-2-yl; ONB, *o*-nitrobenzyl; Mby, *p*-methoxybenzyl; TBDMS, t-butyldimethylsilyl; TIPDS, 1,1,3,3-tetraisopropyldisiloxane-1,3-diyl; Tr, trityl; MTr, 4-methoxytrityl; DMTr, 4,4'-dimethoxytrityl; Trac, trityloxyacetyl; Lev, levulinyl; DBMB, *o*-dibromomethylbenzoyl.

D. Protecting Groups for Phosphoesters and the Phosphorylation Method

In the phosphodiester approach, nucleoside monophosphates are activated by condensing reagents. For the preparation of protected oligonucleotides, the terminal phosphate must be blocked with protecting groups which can be removed selectively. In combination with N-2'-O-acyl and 5'-O-monomethoxytrityl protecting groups, aromatic phosphoramidate was used for terminal protection (Ohtsuka et al., 1970).

In the phosphotriester method, protecting groups for the internucleotide phosphate terminal phosphate are required for preparation of protected oligonucleotide blocks. As a protecting group for internucleotide linkages, phenyl derivatives have been investigated using a dimer containing 2'-O-tetrahydropyranyluridine and the o- or p-chlorophenyl group was found to have suitable stability (van Boom et al., 1974). By employing these protecting groups in the presence of acid-labile 2'-O-protecting groups, temporary protecting groups for the terminal phosphate have to be removed other than acid and alkaline. Figure 5 shows some examples for selective removal at temporary protecting groups from trisubstituted phosphate. The 2,2,2-trichloroethyl group can be removed by reduction with zinc (Eckstein, 1973; van Boom and Burgers, 1978). Aromatic phosphoramidate, which was used as a protecting group for the terminal phosphomonoester, has been employed in combination with p-chlorophenyl (Ohtsuka et al., 1978c, 1979) or o-chlorophenyl (Ohtsuka et al., 1982b, 1984) and converted to the phosphodiester by treatment with

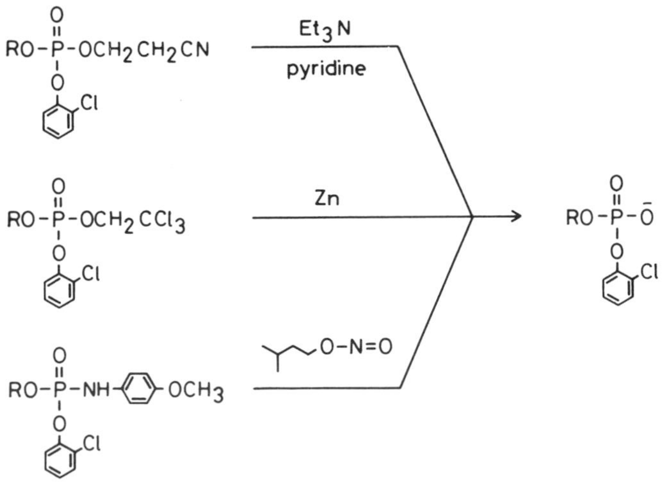

Fig. 5. Selective removal of protecting groups for trisubstituted phosphates.

isoamyl nitrite in a mixture of pyridine and acetic acid. 2,4-Dinitrobenzyl phosphotriesters were found to be stable in pyridine and removed with toluene -*p*-thiol (Christodoulou and Reese, 1983). A stable protecting group, *p*-nitrophenylethyl, can be removed in aprotic solvents by the - elimination mechanism (Uhlmann and Pfleiderer, 1980; Pfleiderer *et al.*, 1980). 5-Chloro-8-quinolyl-*o*-chlorophenyl nucleotides are converted to diester by removing the *o*-chlorophenyl group. The 5-chloro-8-quinolyl group can be removed by chelation with zinc chloride (Takaku *et al.*, 1982). *S,S*-Diphenyl phosphates of ribonucleosides are converted to phosphodiesters by treatment with phosphinic acid–triethylamine (Honda *et al.*, 1984).

E. Condensing Reagents

Dicyclohexylcarbodiimide (DCC) and arenesulfonyl chloride have been discussed in a previous review as activating reagents for phosphomonoesters (Ikehara *et al.*, 1979). Arenesulfonyl azolides have facilitated activation of phosphodiesters as described previously (Reese, 1978; Ohtsuka *et al.*, 1982a; Narang, 1983). The first azolide tested was 1-toluene-*p*-sulfonyl imidazole (Berline *et al.*, 1973). Later, arenesulfonyl triazolides and tetrazolides were prepared and found to activate phosphodiesters rapidly with fewer side reactions than the corresponding chlorides (Stawinski *et al.*, 1977). 1-Mesitylenesulfonyl 3-nitro-1,2,4-triazole has also been prepared and used in the synthesis of oligonucleotides (Reese *et al.*, 1978; Jones *et al.*, 1980). In the presence of a threefold excess of tetrazole arenesulfonyl chlorides activated phosphodiesters as fast as the corresponding tetrazolide (Efimov *et al.*, 1982). This suggests that the sequence of activation of phosphodiester with an arenesulfonyl tetrazolide begins with formation of a mixed anhydride, followed by substitution with tetrazole, and then with formation of a nucleoside (Seth and Jay, 1980) as shown in Fig. 6).

III. SYNTHESIS OF OLIGORIBONUCLEOTIDES BY THE PHOSPHOTRIESTER METHOD

A. Syntheses in Solution

1.   Use of Acid-Labile Protecting Groups for the 2'-OH

For the synthesis of oligoribonucleotides with 3' → 5' internucleotide linkages 2'-O-protected nucleosides are used as starting materials. 2'-*O*-Tetrahydrofuranyl derivatives are used in combination with the 5'-*O*-

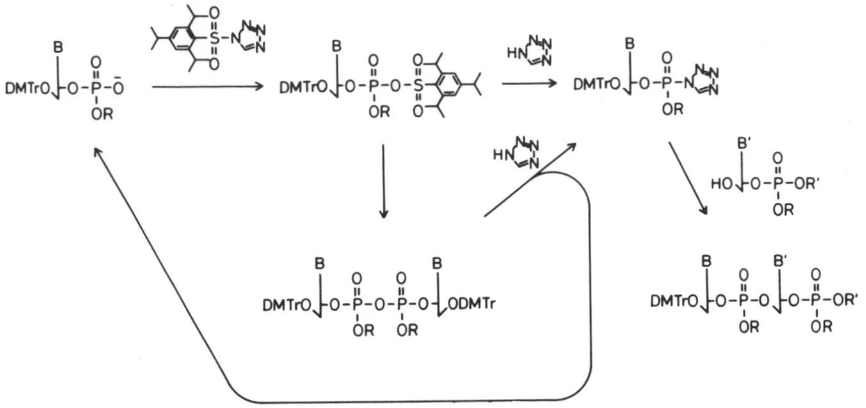

Fig. 6. A mechanism for activation of phosphodiesters with arenesulfonyl azolides.

trityloxyacetyl protecting group for the synthesis of the nonanucleotide corresponding the anticodon loop of tRNA^{Met} (Neilson and Werstiuk, 1974). The 3'-hydroxyl group was phosphorylated with 2,2,2-trichloroethyl phosphate and the nucleotide was condensed with N,-2'-O-protected nucleosides using 2,4,6-triisopropylbenzenesulfonyl chloride (TPS) as the condensing reagent. Later mesitylenesulfonyl triazolide (MST) was used in an improved synthesis (Werstiuk and Neilson, 1976).

The levulinyl group on the 5'-hydroxyl function and the trichloroethyl group on the 3'-phosphate are removed with dilute hydrazine and zinc (van Boom and Burgers, 1975; 1978) in the synthesis of a tetradecaribonucleotide containing repeating UA sequences. The method has been applied to the synthesis of oligoribonucleotides with sequences identical to the nucleation region of tobacco mosaic virus RNA (den Hartog et al., 1981a) as shown in Fig. 7. The same protecting groups were used in the synthesis of an RNA fragment UACGC by the benzotriazolyl phosphotriester approach (Wressman et al., 1983).

Synthesis of the 3'-terminal decanucleotide of yeast tRNA was performed using the 5'-O-(2-dibromomethylbenzoyl) protecting group (Jones et al., 1980) and later 5'-O-2-(methylthiomethoxymethyl)benzoyl groups was introduced for the synthesis involving the 3'-(2,4-dinitrobenzyl) phosphate group (Reese, 1985).

The acid-labile dimethoxytrityl group was used in the presence of 2'-O-tetrahydrofuranyl group and removed by treatment with 2% p-toluenesulfonic acid. By using this procedure, the 3'-terminal nonaribonucleotide of rous sarcoma virus 35 S RNA was synthesized (Takaku et al., 1983).

The 5'-O-dimethoxytrityl group could be removed by treatment with

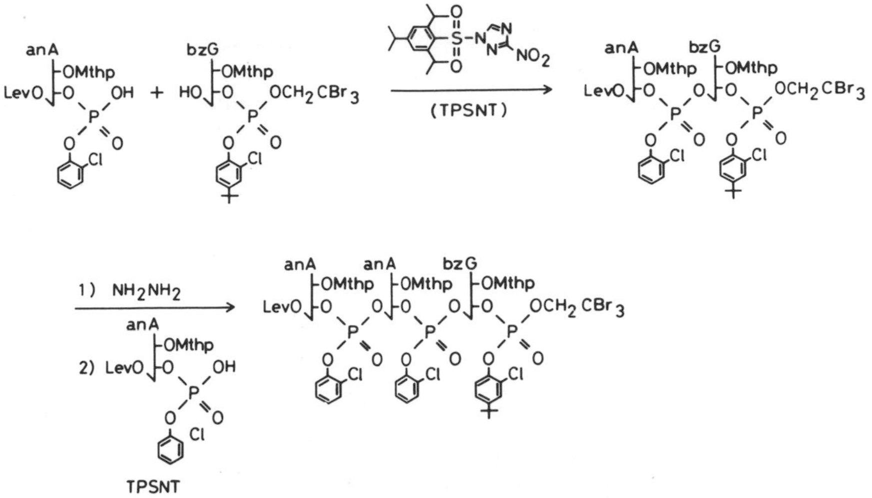

Fig. 7. Synthesis of fully protected oligonucleotide blocks with the 5'-levulinyl group. TPSNT, 1-(2,4,6-Triisopropylbenzenesulfonyl)-3-nitro-1,2,4-triazole.

zinc bromide using conditions as in the deoxyseries (Kierzek *et al.*, 1981). However, the contamination of acids in the reagents has to be avoided in order to maintain the acid-labile protecting group. In combination with the 2'-*O*-tetrahydrofuranyl group, the method has been used for the block synthesis of oligoribonucleotides. The dodecamer containing the termination codon U(AGU)₃AG was synthesized as shown in Fig. 8 (Ohtsuka *et al.*, 1983b) and the rate of removal of the dimethoxytrityl group was found

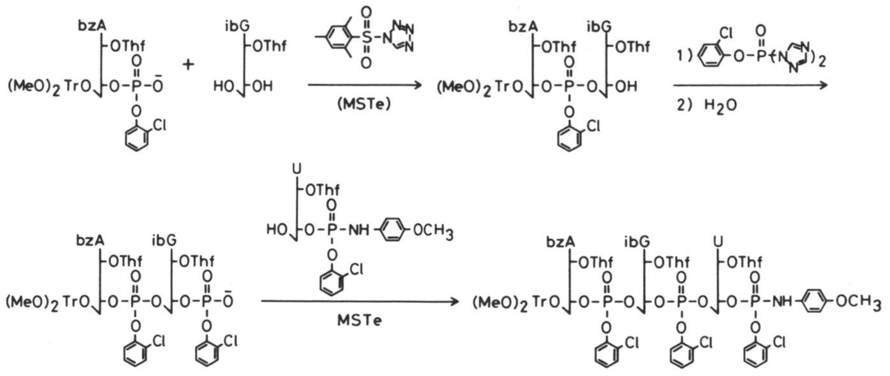

Fig. 8. Synthesis of fully protected oligonucleotide blocks with the 5'-dimethoxytrityl group and 3'-phosphoroanisidate. MSTE, 1-(Mesitylene-2-sulfonyl)-tetrazole.

to differ for each nucleoside. A 33-mer RNA with a sequence identical to the 3'-half of the *Escherichia coli* glycine tRNA has been synthesized using the same approach (Ohtsuka *et al.*, 1984).

A dodecaribonucleotide GUAUCAAUAAUG which has the modified 5'-terminal structure of brome mosaic virus mRNA No. 4 filament has also been synthesized using the 5'-*O*-dimethoxytrityl protecting group (Kamimura *et al.*, 1984).

2.   Use of Other Protecting Groups for the 2'-OH

The photo-labile *o*-nitrobenzyl protecting group has been used for the synthesis of tRNA[Met] fragments up to chain lengths of 16 and 20 (Ohtsuka *et al.*, 1981c,d) as discussed in the previous section.

2'-*O*-(*tert*-Butyldimethylsilyl) nucleosides have been employed in the phosphotriester synthesis (Sadana and Loewen, 1978). Using mesitylene-sulfonyl tetrazolide as the condensing reagents, A_7, A_{11}, and 17 translation control sequences (AAACAUGAGGA and UUACCCAUGU) were synthesized by the modified phosphotriester method as shown in Fig. 9 (Sung and Narang, 1982). 2'-*O*-*tert*Butyldimethylsilyl nucleosides were also used in combination with the *p*-nitrophenylethyl protecting group (Pfleiderer *et al.*, 1980).

3.   Oligonucleotides with 2' → 5' Internucleotide Linkages

The 2',5'-linked oligoadenylates (2-5A) with a chain length of 3–4 have been found in interferon-treated cells and are considered an important mediator for the antiviral action of interferon (Hovanessian and Kerr, 1978; Kerr and Brown, 1978). These unique oligonucleotides with biological activity can be a target for chemical synthesis using 3'-O-protected nucleosides which are obtained as counterparts to the usual 2'-O-protected nucleosides. Almost all protecting groups for the secondary hydroxyl groups discussed in the previous section have been applied to the

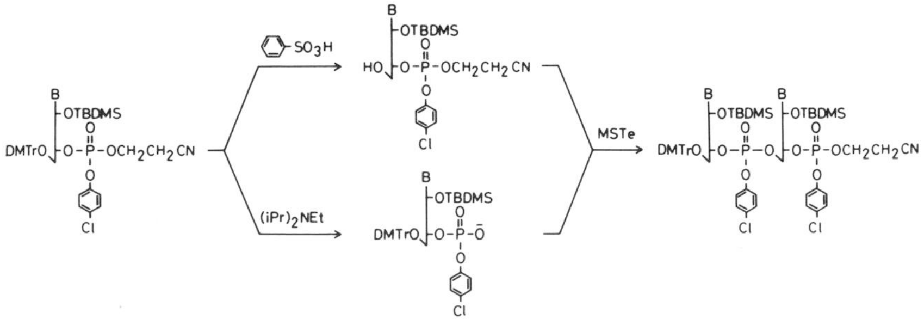

Fig. 9.   Synthesis of fully protected diribonucleotides.

synthesis of 2-5A, and some analogs have also prepared (den Hartog *et al.*, 1981b, and references therein). 2′,5′-Linked core oligonucleotides containing inosine have been prepared using the 2′-*O*-silyl derivative (Charubala and Phleiderer, 1982). Methods of obtaining 2′, 5′-linked oligoadenylate from unprotected nucleosides (Shimidzu *et al.*, 1981) and O-protected nucleosides (Hayakawa *et al.*, 1985) were reported.

B. Solid-Phase Synthesis

In contrast to the polymer-support phosphotriester synthesis of oligodeoxyribonucleotides, solid-phase syntheses of oligoribonucleotides have only been investigated to a limited extent. 2′-*O*-(*o*-Nitrobenzyl) nucleoside derivatives were used in the solid-phase synthesis of a heptaribonucleotides as shown in Fig. 10 (Ohtsuka *et al.*, 1981e). A tridecamer with a repeating sequence has been synthesized using trinucleotides with 2′-*O*-methoxytetrahydropyranyl-5′-*O*-levulinyl protecting groups as shown in Fig. 11 (van der Marel *et al.*, 1982). 2′-*O*-Tetrahydrofuranyl nucleosides have been employed in the polystyrene-support synthesis in combination with the 5′-*O*-dimethoxytrityl group (Ohtsuka *et al.*, 1985). The chain was

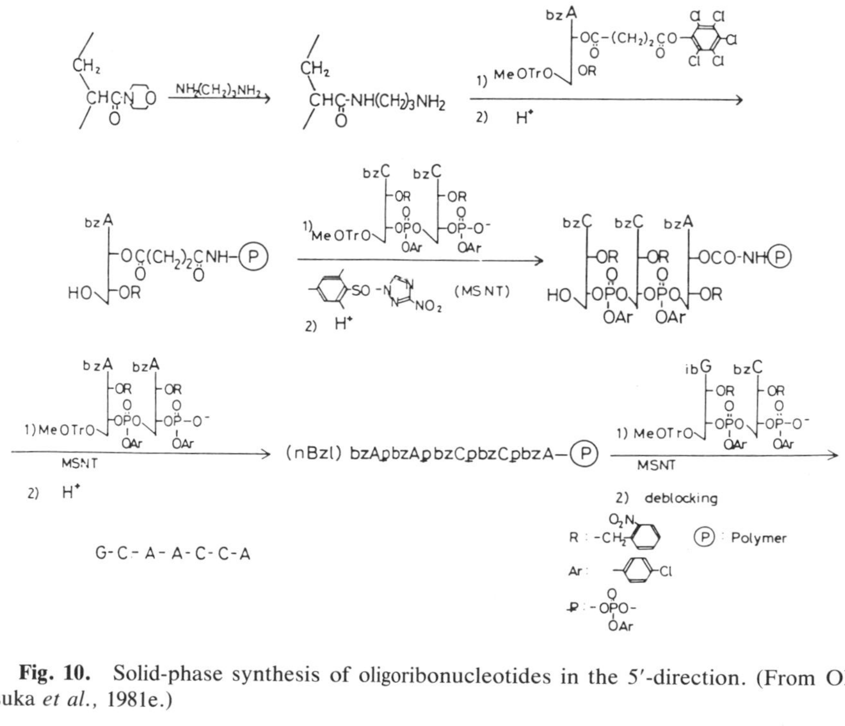

Fig. 10. Solid-phase synthesis of oligoribonucleotides in the 5′-direction. (From Ohtsuka *et al.*, 1981e.)

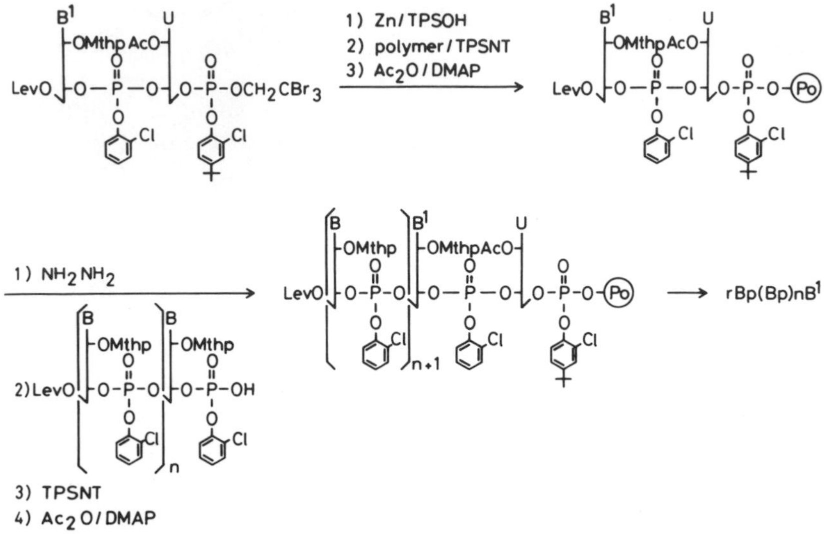

Fig. 11. Solid-phase synthesis of oligoribonucleotides in the 5'-direction starting with a uridine 3'-phosphate. DMAP, 4-Dimethylaminopyridine.

elongated in the 5'-direction by treatment with $ZnBr_2$ using dinucleotides as condensing units. Preparation of protected oligoribonucleotide blocks requires a number of steps due to the protection of the 2'-hydroxyl function. Larger excesses of incoming nucleotides are necessary in the solid-phase synthesis compared to the synthesis in solution. Although, in principle, oligoribonucleotide blocks used in the solution-phase synthesis can be used as condensing units in the solid-phase synthesis, the preparation of protected oligoribonucleotide blocks needs to be simplified. A new approach to the synthesis of oligoribonucleotides in the 3'-direction has been developed by using the 3'-phosphoro-p-anisidate protecting group (Iwai et al., 1985). The chain was elongated by activation of the phosphodiester as illustrated in Fig. 12.

IV. SYNTHESIS OF OLIGORIBONUCLEOTIDES BY THE PHOSPHITE METHOD

A. Synthesis in Solution

Letsinger and co-workers used phosphorochloridites to generate phosphenic triester linkages which can be easily oxidized to the phosphoric

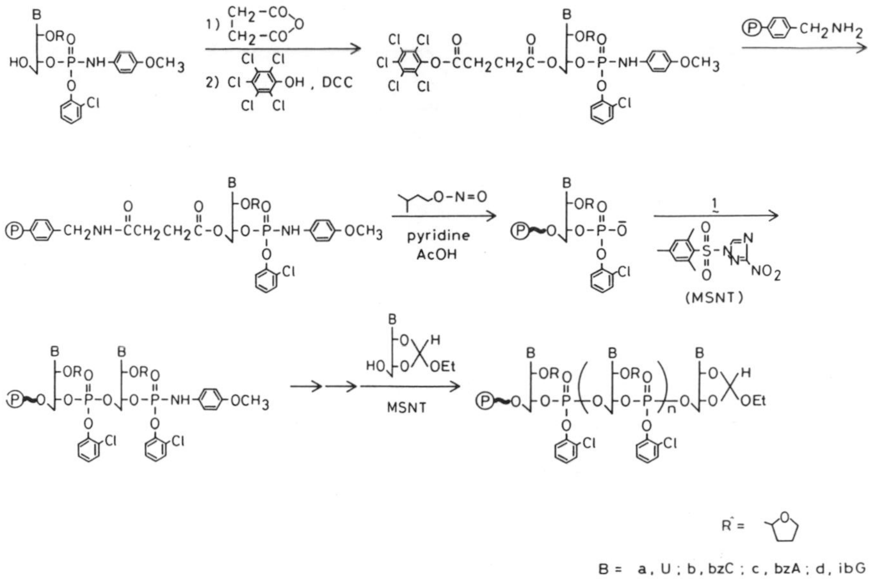

Fig. 12. Solid-phase synthesis of oligoribonucleotides in the 3'-direction.

triesters by treatment with iodine (Letsinger *et al.*, 1975; Letsinger and Lundsford, 1976). The method has been applied to the ribo series (Ogilvie and Theriault, 1979; Ogilvie *et al.*, 1980). N-protected 5'-*O*-monomethoxytrityl-2'-*O*(*tert*-butyldimethylsilyl) nucleosides were phosphitylated with 2,2,2-trichloroethylphosphorodichloridite and used to obtain a heptaribonucleotide GCAACCA by stepwise condensation with a yield ranging from 50 to 87% in each step. Hexadecauridylate was synthesized by using a fully protected diuridylate which was prepared by condensation of 2'-*O*-(*tert*-butyldimethylsilyl)-3'-*O*-levulinyluridine (Ogilvie and Nemer, 1980a) (Fig. 13).

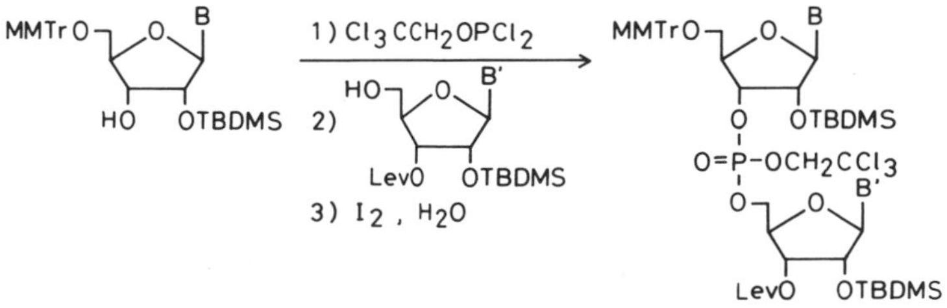

Fig. 13. Phosphite triester synthesis of protected oligoribonucleotides.

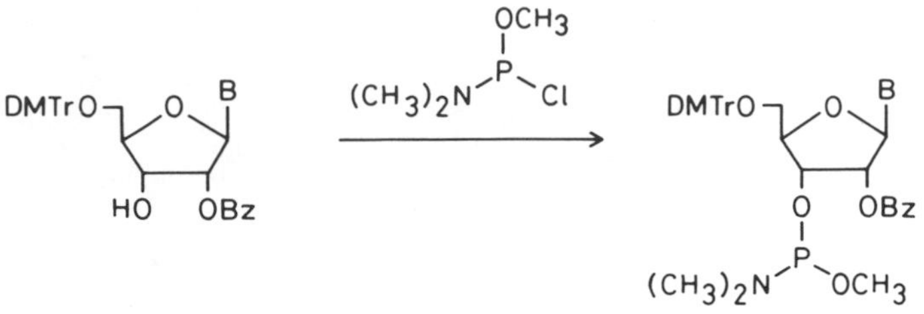

Fig. 14. Preparation of protected nucleoside phosphoramidites.

B. Solid-Phase Synthesis

The phosphite triester method has been employed for the solid-phase synthesis of oligodeoxyribonucleotides and modified by substitution of chlorine for tetrazole or secondary amines (Matteucci and Caruthers, 1981; Beaucage and Caruthers, 1981; Adams et al., 1983) to stabilize intermediates. 5'-O-(Dimethoxytrityl)-2'-O-(benzoyl) or (3,4,5-tri-methoxybenzoyl) base-protected ribonucleosides have been prepared to synthesize oligoribonucleotides on a silica gel support (Kempe et al., 1982) as shown in Fig. 14. Methyl dichlorophosphite was used for the synthesis of oligoribonucleotides on a silica gel (Ogilvie and Nemer, 1980b, 1981). The same procedure has been applied to a mechanical synthesis of oligonucleotides (Ap)$_7$A, (Cp)$_7$C, (Gp)$_7$G, and (Up)$_7$U (Pon and Ogilvie, 1984) and a stepwise synthesis of a nonadecamer corresponding to the tRNA^{Met} fragment (Ogilvie et al., 1984).

V. ENZYMATIC JOINING OF OLIGORIBONUCLEOTIDES BY RNA LIGASE

RNA ligase which can join single-stranded oligoribonucleotides has been isolated from phage T4-infected E. coli (Silber et al., 1972). The enzyme uses ATP as the cofactor (Cranston et al., 1974) and 5'-adenylated oligonucleotides were identified as activated intermediates (Ohtsuka et al., 1976). A comprehensive review article on this enzyme has been published with early references (Gumport and Uhlenbeck, 1981) and some experimental details on ligase reactions have been described in a laboratory manual (Uhlenbeck, 1982; Ohtsuka and Eckert, 1982). Nucleoside 3',5'-diphosphates can be recognized as the smallest substrates and this reaction provides a convenient method for 3'-labeling (England and Uhlenbeck, 1978). Using RNA ligase, single-strand oligoribonucleotides can be joined to produce larger molecules. The nascent formyl-

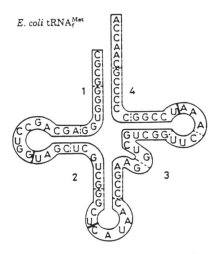

Fig. 15.   Synthesis of *E. coli* formylmethionine-tRNA nascent strand.

methionine-tRNA$_f^{Met}$ of *E. coli* has been synthesized by joining chemically synthesized oligoribonucleotides (Ohtsuka *et al.*, 1981a) (Fig. 15)
and yeast alanine-tRNA has been synthesized by joining small oligonucleotides obtained by chemical or enzymatic procedures (Wang, 1984).
Figure 16 shows the joining reaction for a preparation of the quarter
molecule of tRNA$_f^{Met}$ (section 3 in Fig. 15). Although certain chain lengths
and base sequences are preferred (Ohtsuka *et al.*, 1978b, 1980; Romaniuk
et al., 1982) in the ligase reaction, oligomers with a chain length of up to
100 can be good substrates for RNA ligase.

The joining reaction has been applied to the modification of tRNA
molecules. tRNA$_f^{Met}$ from *E. coli* can be cleaved by RNase A as shown by
arrows in Fig. 17. By combination of this cleavage reaction, replacements
with anticodon triplets, for example with CUA which is complementary
to a nonsense codon UAG, have been performed (Ohtsuka *et al.*, 1983c).

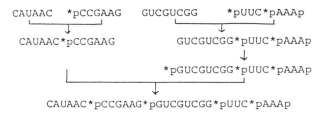

Fig. 16.   Joining of oligoribonucleotides by RNA ligase to construct a quarter molecule
of tRNA.

E. coli tRNA$_f^{Met}$

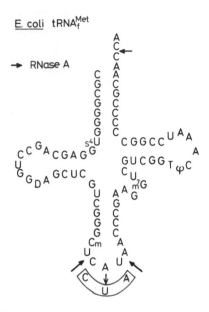

Fig. 17.   Partial hydrolysis with RNase A and anticodon replacement by RNA ligase.

The wobble position of the anticodon in *E. coli* tRNA$_f^{Met}$, has been changed by using RNA ligase (Schulman *et al.*, 1983) and it was demonstrated that these modifications inhibit aminoacylation of *E. coli* tRNA$_f^{Met}$. The effect of the size of the anticodon loop of tRNA on aminoacylation has been investigated by replacement of the anticodon triplet (Schulman and Pelka, 1983; Doi *et al.*, 1985). Modifications of the anticodon loop of yeast phenylalanine-tRNA (Bruce and Uhlenbeck, 1982) and tyrosine-tRNA (Bare and Uhlenbeck, 1985) have been performed by ligation of oligonucleotides. RNA ligase has also been used to construct model substrates for a ribosomal RNA maturation endonuclease (Meyhack *et al.*, 1978).

REFERENCES

Adams, S. P., Kauka, K. S., Wykes, E. J., Holder, S. B., and Galluppi, G. H. (1983). *J. Am. Chem. Soc.* **105**, 661–663.
Bare, L., and Uhlenbeck, O. C. (1985). *Biochemistry* **24**, 2354–2360.
Bartholomew, D. G., and Broom, A. D. (1975). *J. Chem. Soc., Chem. Commun.*, p. 34.
Beaucage, S. L., and Caruthers, M. H. (1981). *Tetrahedron Lett.* **22**, 1859–1862.
Berlin, Yu. A., Chakhmakhcheva, O. G., Efimov, V. A., Kolosov, M. N., and Korobko, G. (1973). *Tetrahedron Lett.*, pp.1353–1356.

Bruce, A. G., and Uhlenbeck, O. C. (1982). *Biochemistry* **21**, 855–861.

Charubala, R., and Pfleiderer, W. (1982). *Tetrahedron Lett.* **23**, 4789–4792.

Chattopadhyaya, J. B., and Reese, C. B. (1978). *J. Chem. Soc., Chem. Commun.*, pp. 639–640.

Chattopadhyaya, J. B., Reese, C. B., and Todd, A. H. (1979). *J. Chem. Soc., Chem. Commun.*, pp. 987–988.

Chladek, S., Zemlicka, J., and Sorm, F. (1966). *Collect. Czech. Chem. Commun.* **31**, 1785–1802.

Christodoulou, C., and Reese, C. B. (1983). *Tetrahedron Lett.* **24**, 951–954.

Cranston, J. W., Silver, R., Malathi, V. G., and Hurwitz, J. (1974). *J. Biol. Chem.* **249**, 7447–7456.

den Hartog, J. A. J., Wilk, G., and van Boom, J. H. (1981a). *Recl. Trav. Chim. Pays-Bas* **100**, 320–330.

den Hartog, J. A. J., Wijmands, R. A., and van Boom, J. H. (1981b). *J. Org. Chem.* **46**, 2242–2251.

Doi, T., Yamane, A., Matsugi, J., Ohtsuka, E., and Ikehara, M. (1985). *Nucleic Acids Res.* **13**, 3685–3697.

Eckstein, F. (1973). In "Protecting Groups in Organic Chemistry" (J. F. W. McOmic, ed.), p. 217. Plenum, New York.

Efimov, V. A., Reverdatto, S. V., and Chakhmakhcheva, O. G. (1982). *Tetrahedron Lett.* **23**, 961–964.

England, T. E., and Uhlenbeck, O. C. (1978). *Nature (London)* **275**, 561–562.

Fromageot, H. P. M., Griffin, B. E., Reese, C. B., and Sulston, J. E. (1967). *Tetrahedron* **23**, 2315–2331.

Gilham, P. T., and Khorana, H. G. (1958). *J. Am. Chem. Soc.* **80**, 6212–6222.

Griffin, B. E., Jarman, M., Reese, C. B., Sulston, J. E., and Trentham, D. R. (1966). *Biochemistry* **5**, 3638–3649.

Griffin, B. E., Jarman, M., Reese, C. B., and Sulston, J. E. (1967). *Tetrahedron* **23**, 2304–2313.

Gumport, R. I., and Uhlenbeck, O. C. (1981). *In* "Gene Amplification and Analysis" (J. G. Chirikjian and T. S. Papas, eds.), vol. 2, pp. 313–345. Elsevier/North Holland, Amsterdam.

Hall, R. H. (1964). *Biochemistry* **3**, 769–773.

Hayakawa, Y., Uchiyama, M., Nobori, T., and Noyori, R. (1985). *Tetrahedron Lett.* **26**, 761–764.

Himmelsbach, F., Schulz, B. S., Trichtinger, T., Charibala, R., and Pfleiderer, W. (1984). *Tetrahedron* **40**, 59–72.

Honda, S., Urakami, K., Koura, K., Terada, K., Sato, Y., Kohno, K., Sekine, M., and Hata, T. (1984). *Tetrahedron* **40**, 153–163.

Hovanessian, A. G., and Kerr, I. M. (1978). *Eur. J. Biochem.* **84**, 149–159.

Ikehara, M., Ohtsuka, E., and Markham, A. F. (1979). *Adv. Carbohydr. Chem. Biochem.* **36**, 135–213.

Iwai, S., Asaka, M., Inoue, H., and Ohtsuka, E. (1985). *Chem. Pharm. Bull.* **33**, 4618–4620.

Jones, S. S., and Reese, C. B. (1979). *J. Chem. Soc., Perkin Trans. I*, pp. 2762–2769.

Jones, S. S., Rayner, B., Reese, C. B., Ubasawa, A., and Ubasawa M. (1980). *Tetrahedron* **36**, 3075–3085.

Jones, S. S., Reese, C. B., Shibanda, S., and Ubasawa, A. (1981). *Tetrahedron Lett.* **22**, 4755–4758.

Kamimura, T., Tsuchiya, M., Urakami, K., Koura, K., Sekine, M., Shinozaki, K., Miura, K., and Hata T. (1984). *J. Am. Chem. Soc.* **106**, 4552–4557.

Kempe, T., Chow, F., Sundquist, W. I., Nardi, T. J., Paulson, B., and Peterson, S. M. (1982). *Nucleic Acids Res.* **10**, 6695–6714.

Kerr, I. M., and Brown, R. E. (1978). *Proc. Natl. Acad. Sci. U.S.A.* **75**, 256–260.

Khorana, H. G. (1968). *Pure Appl. Chem.* **17**, 349–381.

Kierzek, R., Ito, H., Blatt, R., and Itakura, K. (1981). *Tetrahedron Lett.* **22**, 3761–3764.

Köster, H., Kulikowski, K., Liese, T., Heikens, W., and Kohli, V. (1981). *Tetrahedron* **37**, 363–369.

Letsinger, R. L., and Lundsford, W. B. (1976). *J. Am. Chem. Soc.* **98**, 3655-3661.

Letsinger, R. L., Caruthers, M. H., Miller, P. S., and Ogilvie, K. K. (1967). *J. Am. Chem. Soc.* **89**, 7146–7147.

Letsinger, R. L., Finnan, J. L., Heavner, G. A., and Lundsford, W. B. (1975). *J. Am. Chem. Soc.* **97**, 3278–3279.

Lohrmann, R., and Khorana, H. G. (1964). *J. Am. Chem. Soc.* **86**, 4188–4194.

Lohrmann, R., Söll, D., Hagatsu, H., Ohtsuka, E., and Khorana, H. G. (1966). *J. Am. Chem. Soc.* **88**, 819–829.

McGee, D. P. C., Martin, J. C., and Webb, A. S. (1983). *Synthesis,* pp. 540–541.

Markiewicz, W. T. (1979). *J. Chem. Res., Synop.,* p.24.

Markiewicz, W. T., Biala, E., Adamiak, R. W., Kierzek, R., Kysezewski, A., and Wiewirowski (1980). *Nucleic Acids Res., Symp. Ser.* **7**, 115–127.

Matsuzaki, J., Hotoda, H., Sekine, M., and Hata, T. (1984). *Tetrahedron Lett.* **25**, 4019–4022.

Matteucci, M. D., and Caruthers, M. H. (1980). *Tetrahedron Lett.* **21**, 3243–3246.

Matteucci, M. D., and Caruthers, M. H. (1981). *J. Am. Chem. Soc.* **103**, 3185–3191.

Meyhack, B., Puce, B., Uhlenbeck, O. C., and Puce, N. R. (1978). *Proc. Natl. Acad. Sci. U.S.A.* **75**, 3045–3049.

Narang, S. A. (1983). *Tetrahedron* **39**, 3–22.

Neilson, T., and Werstiuk, E. S. (1974). *J. Am. Chem. Soc.* **96**, 2295–2297.

Neilson, T., Deugan, K. V., England, T. E., and Werstiuk, E. S. (1975). *Can. J. Chem.* **53**, 1093–1098.

Norman, D. G., Reese, C. B., and Serafinowska, H. T. (1984). *Tetrahedron Lett.* **25**, 3015–3018.

Ogilvie, K. K. (1983). *In* "Nucleosides, Nucleotides, and Their Biological Applications" (J. L. Rideout, D. W. Henry III, and L. M. Beachma, eds.). Academic Press, New York.

Ogilvie, K. K., and Nemer, M. J. (1980a). *Can. J. Chem.* **58**, 1389–1397.

Ogilvie, K. K., and Nemer, M. J. (1980b). *Tetrahedron Lett.* **21**, 4159–4162.

Ogilvie, K. K., and Nemer, M. J. (1981). *Tetrahedron Lett.* **22**, 2531–2532.

Ogilvie, K. K., and Theriault, N. Y. (1979). *Can. J. Chem.* **57**, 3140–3144.

Ogilvie, K. K., Theriault, N. Y., Seifert, J. M., Pon, R. T., and Nemer, M. J. (1980). *Can. J. Chem.* **58**, 2686-2693.

Ogilvie, K. K., Nemer, M. J., and Gillen, M. F. (1984). *Tetrahedron Lett.* **25**, 1669–1672.

Ohtsuka, E., and Eckert, V. (1982). *In* "Chemical and Enzymatic Synthesis of Gene Fragments" (H. G. Gassen and A. Lang, eds.), pp. 169–178.

Ohtsuka, E., Murao, K., Ubasawa, M., and Ikehara, M. (1970). *J. Am. Chem. Soc.* **92**, 3441–3445.

Ohtsuka, E., Tanaka, S., and Ikehara, M. (1974). *Nucleic Acids Res.* **1**, 1351–1357.

Ohtsuka, E., Nishikawa, S., Sugiura, M., Ikehara, M. (1976). *Nucleic Acids Res.* **3**, 1613–1623.

Ohtsuka, E., Tanaka, S., and Ikehara, M. (1977a). *Chem. Pharm. Bull.* **25**, 949–959.

Ohtsuka, E., Tanaka, S., and Ikehara, M. (1977b). *Synthesis,* pp. 453–454.

Ohtsuka, E., Nakagawa, E., Tanaka, T., Markham, A. F., and Ikehara, M. (1978a). *Chem. Pharm. Bull.* **26,** 2998–3006.
Ohtsuka, E., Nishikawa, S., Markham, A. F., Tanaka, S., Miyake, T., Wakabayashi, T., Ikehara, M., and Sugiura, M. (1978b). *Biochemistry* **17,** 4894–4899.
Ohtsuka, E., Tanaka, T., Tanaka, S., and Ikehara, M. (1978c). *J. Am. Chem. Soc.* **100,** 4580–4584.
Ohtsuka, E., Tanaka, T., and Ikehara M. (1979). *J. Am. Chem. Soc.* **101,** 6409–6414.
Ohtsuka, E., Doi, T., Uemura, H., Taniyama, Y., and Ikehara, M. (1980). *Nucleic Acids Res.* **8,** 3909–3916.
Ohtsuka, E., Tanaka, S., Tanaka, T., Miyke, T., Markham, A. F., Nakagawa, E., Wakabayashi, T., Taniyama, Y., Nishikawa, S., Fukumoto, R., Uemura, H., Doi, T., Tokunaga, T., and Ikehara, M. (1981a). *Proc. Natl. Acad. Sci. U.S.A.* **78,** 5493–5497.
Ohtsuka, E., Wakabayashi, T., Tanaka, S., Tanaka, T., Oshie, K., Hasegawa, A., and Ikehara, M. (1981b). *Chem. Pharm. Bull.* **29,** 318–324.
Ohtsuka, E., Fujiyama, K., Tanaka, T., and Ikehara, M. (1981c). *Chem. Pharm. Bull.* **29,** 2799–2806.
Ohtsuka, E., Fujiyama, K., and Ikehara, M. (1981d). *Nucleic Acids Res.* **9,** 3505–3522.
Ohtsuka, E., Takashima, H., and Ikehara M. (1981e). *Tetrahedron Lett.* **22,** 765–768.
Ohtsuka, E., Taniyama, Y., Marumoto, R., Sato, H., Hirosaki, H., and Ikehara, M. (1982a). *Nucleic Acids Res.* **10,** 2597–2608.
Ohtsuka, E., Ikehara, M., and Söll, D. (1982b). *Nucleic Acids Res.* **10,** 6553–6570.
Ohtsuka, E., Ohkubo, M., Yamane, A., and Ikehara, M. (1983a). *Chem. Pharm. Bull.* **31,** 1910–1916.
Ohtsuka, E., Yamane, A., and Ikehara, M. (1983b). *Nucleic Acids Res.* **11,** 1325–1335.
Ohtsuka, E., Doi, T., Fukumoto, R., Matsugi, J., and Ikehara, M. (1983c). *Nucleic Acids Res.* **11,** 3863–3872.
Ohtsuka, E., Yamane, A., and Ikehara, M. (1984). *Tetrahedron* **40,** 47–57.
Ohtsuka, E., Matsugi, J., Yamane, A., Morioka, H., and Ikehara, M. (1985). *Chem. Pharm. Bull.* **33,** 4152–4159.
Oikawa, Y., Yoshioka, T., and Yonemitsu, O. (1982). *Tetrahedron Lett.* **23,** 885–888.
Pfleiderer, W., Uhlmann, E., Charubala, R., Flockerzi, D., Siber, G., and Verma, R. S. (1980). *Nucleic Acids Symp. Ser.* **7,** 61–71.
Pon, R. T., and Ogilvie, K. K. (1984). *Tetrahedron Lett.* **25,** 713–716.
Ralph, R. K., and Khorana, H. G. (1961). *J. Am. Chem. Soc.* **83,** 2926–2934.
Rammler, D. H., and Khorana, H. G. (1962). *J. Am. Chem. Soc.* **84,** 3112–3122.
Reese, C. B. (1978). *Tetrahedron* **34,** 3143–3179.
Reese, C. B. (1985). *Nucleosides Nucleotides* **4,** 117–127.
Reese, C. B., and Skone, P. A. (1984). *J. Chem. Soc., Perkin Trans. 1,* pp. 1263–1271.
Reese, C. B., and Trentham, D. R. (1965). *Tetrahedron Lett.,* pp. 2467–2470.
Reese, C. B., and Ubasawa, A. (1980a). *Tetrahedron Lett.* **21,** 2265–2268.
Reese, C. B., and Ubasawa, A. (1980b). *Nucleic Acids Symp. Ser.* **1,** 5–21.
Reese, C. B., Suffhill, R., and Sulston, J. E. (1967). *J. Am. Chem. Soc.* **89,** 3366–3368.
Reese, C. B., Stewart, J. C. M., van Boom, J. H., de Leeuw, H. P. M., Nagel, J., and de Rooy, J. F. M. (1975). *J. Chem. Soc., Perkin Trans. 1,* pp. 934–936.
Reese, C. B. Titmus, R., and Yau, L. (1978). *Tetrahedron Lett.,* pp. 2727–2730.
Romaniuk, E., McLaughlin, L. W., Neilson, T., and Romaniuk, P. J. (1982). *Eur. J. Biochem.* **125,** 639–643.
Sadana, K. L., and Loewen, P. C. (1978). *Tetrahedron Lett.,* pp. 5095–5098.
Sasaki, T., and Mizuno, Y. (1967). *Chem. Pharm. Bull.* **15,** 894–896.

Schulman, L. H., and Pelka, H. (1983). *Proc. Natl. Acad. Sci. U.S.A.* **80,** 6755–6759.
Schulman, L. H., Pelka, H., and Susami, M. (1983). *Nucleic Acids Res.* **11,** 1439–1455.
Seth, A. K., and Jay, E. (1980). *Nucleic Acids Res.* **8,** 5445–5459.
Shimidzu, T., and Letsinger, R. L. (1968). *J. Org. Chem.* **33,** 708–711.
Shimidzu, T., Yamana, K., Murakami, A., and Nakamichi, K. (1981). *Tetrahedron Lett.* **22,** 2717–2720.
Silber, R., Malati, V. G., and Hurwitz, J. (1972). *Proc. Natl. Acad. Sci. U.S.A.* **69,** 3007–3013.
Smith, M., Rammber, D. H., Goldberg, I. H., and Khorana, H. G. (1962). *J. Am. Chem. Soc.* **84,** 430–440.
Stawinski, J., Hozumi, T., Narang, S. A., Bahl, C. B., and Wu, R. (1977). *Nucleic Acids Res.* **4,** 353–371.
Sung, W. L., and Narang, S. A. (1982). *Can. J. Chem.* **60,** 111–120.
Takaku, H., and Kamaike, K. (1982). *Chem. Lett.,* pp. 189–192.
Takaku, H., Kamaike, K., and Kasuga, K. (1982). *J. Org. Chem.* **47,** 4937–4940.
Takaku, H., Ueda, S., and Ito, T. (1983). *Tetrahedron Lett.* **24,** 5363–5366.
Takaku, H., Ueda, S., and Tomita, Y. (1984). *Chem. Pharm. Bull.* **32,** 2882–2885.
Ti, C. S., Goffney, and Jones, R. A. (1982). *J. Am. Chem. Soc.* **104,** 1316–1319.
Uhlenbeck, O. C. (1982). *In* "Chemical and Enzymatic Symthesis of Gene Fragments" (H. G. Gussen and A. Lang, eds.), pp. 161–168. Verlag Chemie, Weinheim.
Uhlmann, E., and Pfleiderer, W. (1980). *Tetrahedron Lett.,* pp. 1187–1184.
van Boom, J. H., and Burgers, P. M. J. (1975). *Tetrahedron Lett.,* pp. 4875–4878.
van Boom, J. H., and Burgers, P. M. J. (1978). *Recl. Trav. Chim. Pays-Bas* **97,** 73–80.
van Boom, J. H., Burgers, P. M. J., van Daursen, P.H., Arenzen, R., and Reese, C. B. (1974). *Tetrahedron Lett.,* pp. 3785–3788.
van der Marel, G. A., Wille, G., and van Boom, J. H. (1982). *Recl. Trav. Chim. Pays-Bas* **101,** 241–246.
Wang, Y. (1984). *Acc. Chem. Res.* 393–397.
Watanabe, K. A., and Fox, J. J. (1966). *Angew. Chem., Int. Ed. Engl.* **5,** 579–580.
Watkins, B. E., Kiely, J. S., and Rapoport (1982). *J. Am. Chem. Soc.* **104,** 5702–5708.
Weber, H., and Khorana, H. G. (1972). *J. Mol. Biol.* **72,** 219–249.
Welch, C. T., and Chattopadhyaya, J. (1983). *Acta Cham. Scand. Ser. B.* **B37,** 147–150.
Werstiuk, E. S., and Nielson, T. (1972). *Can. J. Chem.* **50,** 1283–1291.
Werstiuk, E. S., and Nielson, T. (1976). *Can. J. Chem.* **54,** 2689–2696.
Wressmann, C. T. J., Fidder, A., van der Marel, G. A., and van Boom, J. H. (1983). *Nucleic Acids Res.* **11,** 8389–8405.
Zemlicka, J., and Holy, A. (1967). *Collect. Czech. Chem. Commun.* **32,** 3159.

6

Rapid DNA Sequence Analysis

RAY WU
Department of Biochemistry
Section of Molecular and Cell Biology
Cornell University
Ithaca, New York 14853

ROBERT YANG
Section of Molecular Genetics
Division of Biological Sciences
National Research Council of Canada
Ottawa, Ontario K1A OR6, Canada

I. INTRODUCTION

DNA is the genetic material of all living organisms. Information encoded by the sequence of bases in DNA molecules determines the characteristics of each cell by directing protein synthesis and regulating gene expression. The genome of even a simple eukaryotic cell contains over a million base pairs. The genome size of human beings, other mammals, and higher plants exceeds one billion base pairs. Thus, developing methods for rapid DNA sequence analysis is absolutely essential for decoding the vast amount of information stored in the genetic material.

Between 1975 and 1977, two relatively rapid DNA sequencing methods were developed. They include the dideoxynucleotide chain-termination method of Sanger and Coulson (1975) and Sanger *et al.* (1977) and the chemical modification method of Maxam and Gilbert (1977). At that time, these methods were faster than earlier methods developed between 1970 and 1975 (Wu and Taylor, 1971; Murray and Murray, 1973; Sanger *et al.*, 1973) by a factor of more than 10. Between 1977 and 1982, the speed of DNA sequencing increased again by a factor of 10 due to several improvements in the original rapid sequencing methods (for a review, see Wu, 1978; Hindley, 1983; Wu *et al.*, 1983). In this chapter, we summarize the various rapid methods which were published between 1975 and 1985.

Synthesis and Applications
of DNA and RNA

II. ORIGINAL RAPID DNA SEQUENCING METHODS DEVELOPED BETWEEN 1975 AND 1977

A. The Plus-and-Minus Method

Method 1: The plus-and-minus method and the dideoxynucleotide chain-termination method.

The first rapid method for determining nucleotide sequences in DNA was developed by Sanger and Coulson (1975) and is known as the plus-and-minus method. This method is based on primer-dependent DNA synthesis catalyzed by DNA polymerase to generate labeled DNA products of different lengths, which are then fractionated on polyacrylamide gels. The success of this method depends on the elegant design by Sanger and Coulson (1975) to generate DNA of every length and the resolving power of electrophoresis on thin gels to separate a family of DNA products whose members differ in length by a single nucleotide. The sequence of the DNA can be read directly from the autoradiogram of the gel.

In the plus-and-minus method, on strand of the DNA of interest serves as a template (Fig. 1A). By adding DNA polymerase and four deoxynucleoside triphosphates (dNTPs), one of which is ^{32}P-labeled, complementary strands of different lengths are synthesized (Fig. 1B).

1. The Minus System

In the *minus system,* the labeled complementary strand serves as a primer. Primer extension is carried out according to the partial-repair principle of Wu and Kaiser (1968), who also introduced the method of labeling DNA using DNA polymerase I as the basic approach for sequencing DNA. Synthesis proceeds as far as it can on each chain and stops when positions are reached where the missing dNTP is required. If dATP is the missing triphosphate (the minus A system), each chain will terminate at a position before a dA residue (Fig. 1C, left-handed panel). Four separate samples are incubated, with one of the four dNTPs missing in each sample. The DNA products from the four incubation mixtures are then denatured and fractionated by gel electrophoresis.

2. The Plus System

In the *plus system,* the method of Englund (1972) is used. In the presence of a single dNTP, T4 DNA polymerase degrades double-stranded DNA from its 3′ ends and stops at residues corresponding to the single dNTP that is present due to the incorporation of this added nucleotide. This method is coupled to the use of the random DNA mixture shown in

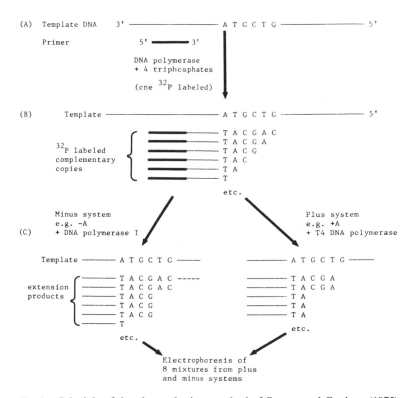

Fig. 1.   Principle of the plus-and-minus method of Sanger and Coulson (1975).

Fig. 1B. DNA samples are incubated with T4 DNA polymerase and one of the four dNTPs. For example, in a plus A system, all chains will terminate with a dA (Fig. 1C, right-hand panel). Four separate samples are incubated, with one of the dNTPs added in each sample. The DNA products (Fig. 1C) are then fractionated by gel electrophoresis.

B. The Dideoxynucleotide Chain-Termination Method

The plus-and-minus method was improved by Sanger *et al.* (1977) by introducing dideoxynucleoside triphosphates (ddNTPs) as DNA chain terminators. As shown in Fig. 2, a primer (decanucleotide) is incubated with a template DNA in the presence of *Escherichia coli* DNA polymerase (polymerase I) and four dNTPs (one of them ^{32}P-labeled) plus a ddNTP (here, ddTTP is used). During the primer extension reaction, at any position where a dTTP is needed, either dTMP or a ddTMP (ddT) is incorporated. Wherever a ddT is incorporated, the DNA chain is termi-

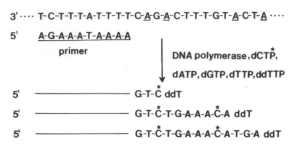

Fig. 2. Principle of the dideoxynucleotide chain-termination method for sequencing DNA (Sanger *et al.*, 1977). The upper line gives a template sequence of a single-stranded DNA. The short sequence immediately below represents a primer, which is complementary to a part of the template sequence.

nated. Thus, a family of DNA fragments sharing the same 5' end, but having ddT residues at different locations is generated. By using a different ddNTP in four different samples, four families of DNA fragments that share the same 5' end are generated. The four samples are denatured and fractionated by electrophoresis in a denaturing polyacrylamide gel. The sequence of the extended primer, which is complementary to the template sequence, can be read directly from the autoradiogram of the gel.

C. The Chemical Cleavage Method

Method 2:   The base-specific chemical cleavage method.

A rapid and direct method for sequencing DNA based on chemical reactions that break single-end-labeled DNA molecules at specific nucleotide bases has been developed by Maxam and Gilbert (1977, 1980). This method has become widely used since its introduction in 1977. In the chemical cleavage method, four separate reactions are needed in which one (or two) of the four nucleotide bases is modified; the DNA backbone is then cleaved preferentially at that point. The characteristic features of this method in comparison to the enzymatic one include: (1) no *in vitro* enzyme copying reaction is needed; (2) either single- or double-stranded DNA molecules can be directly sequenced; (3) the terminal labels of DNA fragments to be analyzed can be either at the 3' or the 5' end.

Specific cleavage is central to the chemical method of sequencing DNA. Each sequencing reaction involves a three-step process: modification of a base, removal of the modified base from its sugar, and finally

DNA strand scission at that sugar. In most cases, the modification reaction disrupts the electronic structure of a DNA base, rendering it vulnerable for a second reaction to break the bond between the base and its sugar. For example, dimethyl sulfate methylates the N-7 of guanine in DNA, leaving a positive charge on the N-7–C-8–N-9 imidazole portion of the purine ring. Subsequently, piperidine displaces the 7-methylguanine and catalyzes β-elimination of both the 3'- and 5'-phosphates from the sugar. Although dimethyl sulfate also methylates adenine (at N-3) in the same reaction, piperidine does not react with methylated adenine and no strand scission takes place.

In each designated reaction, a family of subfragments is generated, extending from the same labeled end to the positions of cleavages (for details, see legends to Fig. 3). When the products of the four base-specific reactions are separated by gel electrophoresis, the DNA sequence can be deduced from the radioactive banding patterns displayed on an autoradiogram (Fig. 3).

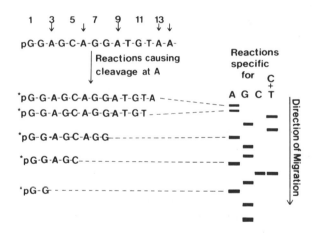

Fig. 3. Principle of the chemical method for sequencing DNA (Maxam and Gilbert, 1977). A 14-nucleotide-long portion of a single-end-labeled DNA molecule is used as an example. The DNA molecules are labeled at the 5' ends with ^{32}P (asterisk), and are subjected to base-specific chemical modification and cleavage reactions. For example, by using the adenine (A)-specific reaction, some molecules are cleaved only at position 14 producing a 13-residue-long radioactive product. Other molecules are cleaved at 13, 9, 6, or 3, yielding 12, 8, 5, or 2-base-long radioactive products, respectively. After gel electrophoresis, the five radioactive products are separated by gel electrophoresis (far left). Three other reactions (specific for G, C, and T) are simultaneously carried out and the mixtures are electrophoresed on the same slab gel. The sequence is then directly read from the autoradiogram starting from the bottom.

III. IMPROVED RAPID METHODS DEVELOPED BETWEEN 1977 AND 1982

A. Sequence Analysis of DNA up to 500 Base Pairs

Method 3:   Sequencing DNA after restriction digestion and shotgun cloning into a M13 phage.

The original enzymatic sequencing method (Method 1) requires that the template DNA is in single-stranded form. Preparation of the single-stranded DNA from a duplex DNA is not always straightforward, and this limits the scope of the procedure. To circumvent this problem and others, Sanger *et al.* (1977, 1980) introduced a method using dideoxynucleotide chain-termination combined with cloning of small fragments of DNA in a single-stranded phage vector. The shortened random fragments derived from restriction enzyme digestion of the DNA are inserted into the *Eco*RI site of M13 mp2 through an oligonucleotide linker. Next, the single-stranded DNA from each phage clone is sequenced from one end using a DNA primer which is complementary to the M13 sequence flanking the insert. Under normal conditions, a sequence of approximately 300 nucleotides can be resolved per fragment using a 40-cm gel.

Because each fragment can be cloned in either orientation, the two complementary sequences of the same fragment can be readily obtained. In this method, selection of phage clones with DNA inserts is simplified by taking advantage of the special features of the modified M13 mp2 phage (Gronenborn and Messing, 1978). This phage was constructed by inserting a portion of *E. coli lac* operon into the intragenic space of the wild-type M13 genome. The inserted segment mainly contains the *lac* promoter, the operator, and the proximal segment of the *lacZ* gene, which codes for the N-terminal 145-amino-acid α-peptide sequence of β-galactosidase. Within the α-peptide gene, a unique *Eco*RI site was installed. When a *E. coli* K12 host, which produces a defective β-galactosidase lacking the sequence of amino acid 11-41 within the α-peptide, is infected by this phage, blue plaques will be generated, due to α-complementation, on an indicator plate containing isopropyl-β-D-thiogalactoside (IPTG, an inducer) and 5-bromo-4-chloro 3-indolyl-β-D-galactoside (X-gal, a chromogenic substrate).

When a foreign DNA is inserted into the *Eco*RI site, synthesis of α-peptide is disrupted. The recombinant phages give rise to the white plaques due to the absence of α-complementation. Improved vectors such as M13 mp18 and 19, which possess as many as 12 unique cloning sites within a short polylinker region, have been constructed by Yanisch-Perron *et al.* (1985).

Similar shotgun strategies using different tools such as sonication (Deininger, 1983) or DNase I (Anderson, 1981) have also been described. In these random shotgun strategies, no detailed physical mapping of the target DNA is needed. However, a large number of sequencing experiments are required to ensure the complete representation of the entire DNA in the assembled sequences. Thus, this method depends heavily upon the use of computer analysis to store and interpret the data. Sequence data can be obtained rapidly at the beginning, but slowly toward the end. Therefore, the time needed to determine the final 10% of the sequence may be equal to that needed to determine the initial 90%.

Method 4: Rapid chemical sequencing methods that do not require prior separation of labeled DNA fragments.

In the original chemical cleavage method, end-labeled DNA fragments must first be isolated by gel elctrophoresis to give single-end-labeled fragments (Maxam and Gilbert, 1977). An improved method was devised by Rüther *et al.* (1981) in which the cloned DNA fragments could be sequenced directly and rapidly since no purification of the labeled DNA by gel electrophoresis was needed. In this method, a muitiple-purpose plasmid, pUR250, which contains several unique cloning sites (e.g., *Hin*dIII, *Xba*I, *Sal*I, *Acc*I, *Hin*cII, *Bam*HI, and *Eco*RI) in a small region of *lacZ* gene locus was constructed. This vector allows cloning of the DNA of interest (foreign DNA) by inserting it into the region of the polylinker sites. A selection system similar to the M13 system is employed in which bacteria containing recombinant plasmids generally give rise to white colonies, while those receiving plasmids without inserts form blue colonies on indicator plates. As shown in Fig. 4, a DNA fragment of interest (hatched region) is inserted into pUR250 at the *Hin*cII site (steps a and b). The recombinant plasmid is cut with *Bam*HI and terminally labeled by incorporating cold dG and hot dA with *E. coli* DNA polymerase I (steps c and d). The both-end-labeled *Bam*HI fragment is cleaved with *Eco*RI (step e) to give two single-end-labeled fragments, a long one with the insert sequence and a short one with only 11 base pairs. The labeled fragments are then subjected to the chemical degradation sequencing procedure (method 2). Because the contaminated short fragment has limited length (11 base pairs), the sequence of the insert DNA can be read without interference. Under normal conditions, as many as 300 bases of the upper strand can be readily obtained. We can also apply the same principle to start from the other end of the insert DNA using *Xba*I to generate ends for labeling (steps c and d) and *Hin*dIII to produce the single-end-labeled fragments (step c). Thus, one strand of a DNA fragment of 500 base pairs in length can be rapidly sequenced. For sequencing a longer DNA frag-

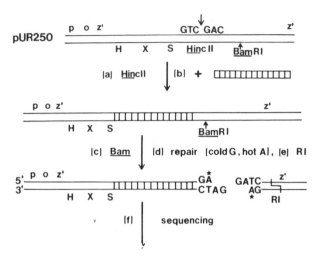

Fig. 4.   Direct sequencing of labeled DNA without prior gel purification.

ment (for example, 1000–3000 base pairs in length), appropriate restriction cleavage is needed to give smaller fragments which are then subcloned into the multipurpose plasmid for direct sequencing.

Method 5:   Use of synthetic primers for sequencing DNA cloned in pBR322.

The use of synthetic oligodeoxynucleotides as primers for DNA sequence analysis was proposed by Wu (1972) and first put into practice by Sanger *et al.* (1973). Several synthetic primers for sequencing DNA cloned into M13 phages were described later (Narang *et al.,* 1980; Messing *et al.,* 1981). The primer is first hybridized to the single-stranded M13 phage DNA adjacent to the DNA fragment to be sequenced. The sequence of the DNA fragment is obtained after extending the primer using the dideoxynucleotide chain-termination method.

Smith *et al.* (1979) used a synthetic primer for sequencing the yeast cytochrome *c* gene cloned in a plasmid. Later, Wallace *et al.* (1981) synthesized seven primers for rapidly obtaining the sequence of DNA clonced in pBR322. In this method, the double-stranded plasmid is linearized and then denatured. An excess amount of a primer (20-fold excess over the template) is added and is allowed to hybridize to the template. Primer extension can give the sequence of up to 300 nucleotides of one strand of DNA cloned into the *Pst*I or *Sal*I site of pBR322, or the sequence of both strands of DNA cloned into the *Bam*HI, *Eco*RI, or *Hind*III site. Although excess primer can compete with the complementary strand of the template DNA, the pattern of the sequencing gel is often less clean than using single-stranded templates.

For the fractionation of DNA fragments on gels, usually a 40-cm-long gel is used; up to 300 base pairs can be obtained with two or three loadings. However, by using a 80-cm-long gel, up to 600 base pairs can be obtained (Yang and Wu, 1979; Smith and Calvo, 1980).

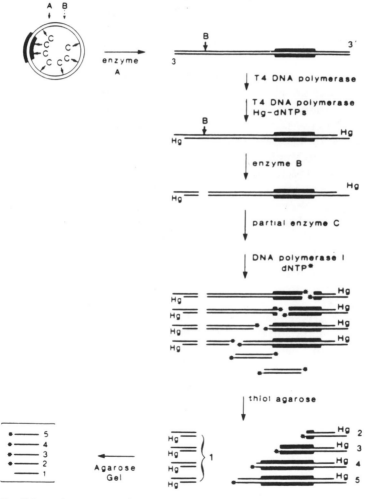

Fig. 5. Schematic representation of the mercury-thiol affinity system for generating labeled overlapping DNA molecules for sequencing (Hartley *et al.,* 1982). The region of DNA to be sequenced is marked with thick lines in the recombinant plasmid. The asterisks represent ^{32}P-labeled nucleotides incorporated into the end of each DNA fragment generated by DNA polymerase I.

B. Sequence Analysis of DNA Longer than 500 Base Pairs

Method 6:   A mercury-thiol affinity column system for DNA sequencing.

Hartley *et al.* (1982) have established an entirely *in vitro* protocol which generates overlapping labeled DNA for direct chemical sequencing so that a long nucleotide sequence can be derived without subcloning. This method requires mercuration of plasmid DNA, purification of DNA on a thiol-agarose affinity column, and end-labeling by DNA polymerase-catalyzed repair synthesis. One preparative gel electrophoresis step yields a number of single-end-labeled fragments which are ready for sequencing reactions.

This method, as schematically illustrated in Fig. 5, involves linearization of the recombinant plasmid by a unique cutting enzyme A, followed by incorporation with mercurated nucleotides using T4 DNA polymerase. After digestion with a second unique cutting enzyme B, the DNA mixture is divided into aliquots which are then partially digested by a multiple-cutting enzyme (enzyme C represents any of the restriction enzymes which produce 3'-recessive ends). Following terminal labeling with [α-^{32}P] dNTPs and DNA polymerase I, the fragments are passed through a thiol-agarose column. The bound mercurated fragments are eluted with 2-mercaptoethanol and applied to a preparative agarose gel electrophoresis. Fragments of interest are eluted and sequenced directly using the chemical cleavage procedure. By overlapping the sequence of the adjacent fragments, the complete sequence of the target DNA fragment of over a thousand nucleotides can be assembled quite rapidly. The same DNA can be approached in the same manner from the opposite end to obtain the complementary sequence data. This method has the advantage over the "shotgun" method in that no subcloning is needed and that the nucleotide sequencing approach is progressive, unidirectional, and systematic.

IV. RAPID METHODS INVOLVING PROGRESSIVE SHORTENING OF DNA

A. Progressive Shortening of Double-Stranded DNA by Enzymes

Progressive shortening of the DNA of interest followed by subcloning of the shortened fragments allows more systematic sequence analysis. This approach is better than the random ("shotgun") method since the sequences from different progressively shortened fragments can be lined up easily without the need of extensive computer analysis. It also avoids the problem of sequencing certain fragments repeatedly, which is typical in the random method.

There are several methods for progressive shortening of the DNA of interest. Although the basic principle is similar, each method makes use of a different enzyme for shortening the DNA and a different strategy for subcloning or analysis of the subclones. Methods 7–10, to be described here, are related in principle to but are simpler than those proposed by Frischauf *et al.* (1980) and Barnes (1980).

Method 7: Progressive shortening of DNA cloned in a M13 phage by DNase I followed by subcloning and sequencing.

Using the replicative form of the M13 phage as the vector, Hong (1982) developed a systematic sequencing strategy in which the DNA fragment to be sequenced (cloned DNA of interest) is progressively shortened. In this method (Fig. 6), the cloned DNA (insert DNA) is digested by restriction enzyme B and ligated to a similarly digested replicative form of M13 DNA to form a recombinant DNA molecule (c). This circular DNA is then digested with a small amount of pancreatic DNase I in the presence of Mn^{2+} and fractionated on an agarose gel. The partially digested linear form of the recombinant DNA (d) is digested at site A with a second restriction enzyme (*Sma*I is used in this case) and treated with DNA polymerase I to create a mixture of products (e) with the DNA of interest (cross-hatched regions) at one end. Circularization of these recombinant DNA molecules with T4 DNA ligase followed by transfection produces a mixture of clones with the insert DNA fixed at one end (near A*) and sequentially shortened at the other end. By using a screening procedure involving an abbreviated dideoxynucleotide chain-termination in the presence of ddTTP (only one reaction per clone), different recombinant clones can be identified in which the insert DNA varies in length by about 200 nucleotides. Sequencing the appropriate clones from the primer site (P) by the dideoxynucleotide chain-termination method (four reactions per clone) can cover the entire length of the original insert DNA. Typically, the overlapping sequences between clones are around 30–50 nucleotides. The advantages of this rapid method are that (1) no additional vector is needed, and (2) no physical mapping of the DNA of interest is necessary.

A modification of Hong's systematic DNA sequencing strategy is described by Lin *et al.* (1985). In this method, after DNase I digestion, the single-cut linear DNA does not have to be separated from the supercoiled or open circular DNA on an agarose gel. After ligation, the DNA is digested with a second restriction enzyme for which a unique cleavage site resides between the insert DNA and the first restriction enzyme cutting size. Thus, the original intact DNA is linearized, whereas the desir-

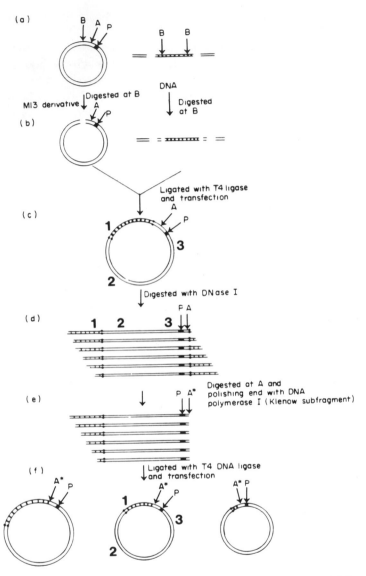

Fig. 6. A scheme for systematic shortening of DNA by pancreatic DNase I followed by subcloning and sequencing (Hong, 1982).

able deleted subclone is not. The background is decreased 25-fold when the ligated DNA mixture is treated with the second restriction enzyme. Moreover, the yield of subclones per microgram of DNA is 60-fold higher in this method than in the original protocol.

Method 8: Progressive shortening of DNA cloned in a M13 phage by nuclease BAL 31 followed by subcloning and sequencing.

A rapid "nonrandom" DNA sequencing procedure, which is suitable for DNA molecules several thousand base pairs in length, was described by Poncz *et al.* (1982). The method involves cloning the DNA to be sequenced into a M13 phage, followed by digesting with a restriction enzyme at a site close to one end of the insert DNA. The resulting linear DNA is digested with nuclease BAL 31 and samples are taken at different time intervals. The progressively shortened DNA fragments are subcloned into a suitably cut M13 vector so that the blunt end is placed next to the primer site. Determination of the size of the insert in the various subclones is accomplished by hybridization to a complementary single-stranded probe derived from M13 containing the total insert followed by nuclease S_1 digestion. A complete DNA sequence of the original DNA of interest can be accomplished by selecting subclones of progressively shorter insert sizes, and by using a DNA primer and the dideoxynucleotide chain-termination method. Thus, analysis of the DNA sequence proceeds from one end of the original insert to the other end in a nonrandom manner. Sequence analysis of both DNA strands is accomplished by construction of two progressive deletion series starting from different ends of the original inserted DNA.

Both Methods 7 and 8 require that the DNA fragment to be sequenced does not contain the same restriction site selected for digestion and manipulation of the recombinant DNA clone.

Method 9: Progressive shortening of DNA cloned in a plasmid by exonucleolytic enzymes followed by subcloning and sequencing.

Exonuclease III, a $3' \rightarrow 5'$ exonucleolytic enzyme that digests double-stranded linear DNA molecules, can convert a duplex DNA into partially single-stranded DNA templates (Richardson *et al.,* 1964). Primers are then added to such templates and sequence analysis can be carried out by primed synthesis (Wu *et al.,* 1973; Smith, 1979; Zain and Roberts, 1979).

Exonuclease III has also been used to shorten the 3' ends of duplex DNA; the shortened 3' ends can serve as primers for direct enzymatic sequencing (Guo and Wu, 1982).

Recently, Guo *et al.* (1983) presented a systematic method for shortening of DNA by digestion with exonuclease III and S_1 nuclease (or BAL 31 nuclease). The shortened DNA of the desired length is selected and subcloned into plasmid pWR2 or pWR34 for direct sequencing. In this method (Fig. 7), the recombinant plasmid p27-34 with a 2.2-kb DNA insert (DNA of interest) at the *Eco*RI site is first linearized by *Hin*dIII which cuts once just outside the insert (step a). Next, progressive digestion with exonuclease III for different lengths of time, followed by cleav-

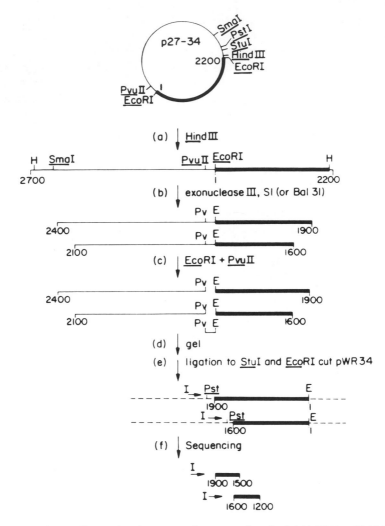

Fig. 7. Diagram illustrating the strategy for sequencing of a 2.2-kb DNA. p27-34 is used as an example to describe the strategy. p27-34 (4.9 kb) contains a 2.2-kb human DNA fragment cloned in the *Eco*RI site of the polylinker region of pWR34. Details for each step are given in the text. Note that between steps (c) and (f), the orientation of the target DNA is reversed for convenience. Abbreviations for restriction enzyme sites are: E, *Eco*RI; Pv, *Pvu*II, and H, *Hin*dIII. The heavy line represents the human DNA fragment to be sequenced, the thin line represents the vector DNA, and the dashed line represents the plasmid pWR34 DNA used for subcloning in step (e).

age with S_1 nuclease (alternatively BAL 31 nuclease can be used) produces a family of shortened duplex DNAS (step b). After cleavage with *Eco*RI and *Pvu*II enzymes (step c), the family of the progressively shortened inserts is isolated by electrophoresis on a low-melting agarose gel. The DNA molecules of interest from different size classes, which differ by 400–500 base pairs, are then ligated (step e) to suitably digested pWR34.

Transformation of *E. coli* is performed and white colonies are picked. The sequence of each recombinant DNA sample is determined by the dideoxynucleotide chain-termination method or the chemical cleavage method (method 4). Since the foregoing method can be applied to either orientation, the complementary sequence of the DNA of interest is readily obtainable. Using this strategy, the entire sequence of a DNA molecule several thousand nucleotides in length can be determined within a short time.

Because exonuclease III shows little base specificity, progressive cleavage using this enzyme can proceed at a relatively uniform rate. However, BAL 31 nuclease digestion tends to pause at G-C-rich regions so that certain parts of the original DNA may not be well represented (Guo and Wu, 1982; Yang *et al.,* 1983).

Another unique property of exonuclease III, observed by R. C. A. Yang and R. Wu (unpublished result, quoted by Guo and Wu, 1982), is that this enzyme only digests DNA with blunt ends or with a 5′ protruding end. DNA molecules with 3′ protruding ends are not digested (or are poorly digested). Based on this property, Henikoff (1984) and Yanisch-Perron *et al.* (1985), recently presented improved methods for sequencing long DNA molecules cloned in M13 using exonuclease III and S_1 nuclease for generating deletion clones of varying lengths.

The usefulness of this method was demonstrated by Henikoff (1984) on a 4570-bp *Drosophila* genomic DNA fragment cloned in the single-stranded phage vector M13 mp18. As shown in Fig. 8, an ordered set of deletion clones was made by first cutting replicative form (RF) DNA with two restriction enzymes (A and B) in the polylinker region between the insert DNA and the sequencing primer binding site. Cutting with one enzyme results in a four-base 3′ protrusion (at the right-hand end) that protects the remainder of the vector from exonuclease III attack. This allows unidirectional (from left to right) and progressive digestion of the insert sequence from the 3′-recessive end produced by the other enzyme. Exonuclease digests from different time intervals were collected, treated with S_1 nuclease, Klenow polymerase, and T4 DNA ligase, and then used to transfect competent *E. coli* cells. The resulting clones were subjected to direct sequencing. This method has one advantage in that it avoids the

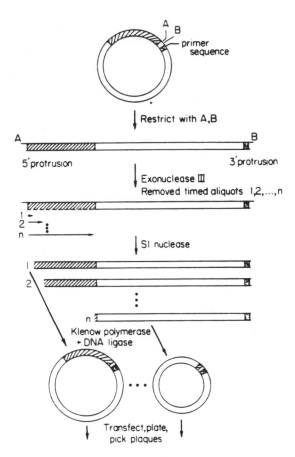

Fig. 8. Outline for the strategy for targeting deletions (Henikoff, 1984). For nucleotide sequencing, aliquots were removed from the exonuclease III reaction and separately processed at different intervals until nearly the entire insert was digested. The DNA of interest (insert DNA) is represented by hatched lines.

problems of cloning shortened fragments into a second vector. Therefore, it is simpler and more rapid than the one proposed by Guo *et al.* (1983).

B. Progressive Shortening of Single-Stranded DNA by Enzymes

Method 10: Progressive shortening of single-stranded M13 DNA carrying the insert by T4 DNA polymerase followed by subcloning and sequencing.

All the methods described so far for the progressive shortening of DNA involves double-stranded DNA. Dale *et al.* (1985) introduced a novel method which involves progressive digestion of single-stranded recombinant M13 DNA. The method starts with an insert DNA fragment cloned in a double-stranded M13 DNA. As shown in Fig. 9 (step a), a single-stranded recombinant M13 DNA is isolated and annealed to a synthetic complementary DNA oligomer (such as a 29-mer which ends with 11 T residues) to form a specific cleavage and ligation substrate. After cleaving the short double-stranded region with a restriction enzyme (such as *Hin*dIII), the linear single-stranded DNA is digested from the 3′ end using the

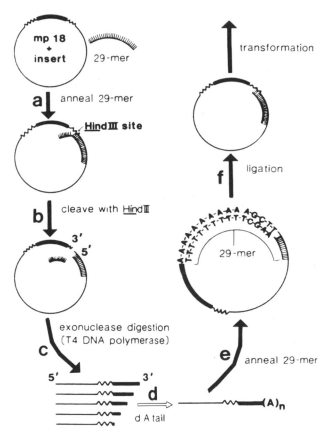

Fig. 9.　Outline of the single-stranded cloning procedure for M13 mp 8, 10, or 18 (Dale *et al.*, 1985). The solid bar represents the insert DNA (DNA to be sequenced) and the sawtooth line the polylinker region of M13. The 29-mer (5′-CGACGGCCAGTGCCAAGCTTTTTTTT-TTTT-3′) is a specially designed 'bandaid' oligomer as mentioned in the text.

$3' \rightarrow 5'$-exonuclease activity T4 DNA polymerase (step c). A short homopolymeric tail (such as a dA tail) is added to each molecule of shortened inserts using terminal transferase (step d) to provide a known $3'$ end that is complementary to the oligomer. The same dT-tailed 29-mer oligomer is next annealed (step e) to the family of shortened and dA-tailed DNA to allow the two ends of each DNA molecule to be ligated (step f). The circular single-stranded DNAs are then used to transform *E. coli* cells. Clones are picked, and the size of insert in each single-stranded recombinant phage is compared by running SDS-disrupted phage on agarose gels. Differences in insert size as small as 124 bases can be detected on 0.7% agarose gels. Suitable single-stranded M13 DNA templates are prepared from clones with the size of inserts differing by approximately 250–350 base pairs. Sequencing reactions are carried out using the dideoxynucleotide chain-termination procedures, and using either the same 29-mer oligomer or a universal primer for M13 as the primer. Dale *et al.* (1985) reported that the subcloning and sequencing in both directions of a 2600-base pair DNA was accomplished by one individual in 5 days.

C. Sequencing DNA Using Single-Stranded Plasmids

Method 11: Sequencing DNA using two new families of single-stranded plasmids.

A single-stranded : double-stranded bifunctional family of pEMBL plasmids has been constructed by Dente *et al.* (1983). The pEMBL plasmids are characterized by the presence of (a) the *bla* gene as selectable antibiotic marker, (b) a short segment coding for α-peptide of β-galactosidase and containing a polylinker with multiple cloning sites, and (c) the intragenic cis-effective region of phage F1 bearing the phage origin of replication. A pEMBL plasmid can produce a single-stranded DNA upon superinfection of the host cells with phage F1. The single-stranded plasmid can be packaged into virion-like particles and excreted in the culture medium. The orientation of the F1 region determines which of the two strands is encapsidated into the viral particles. Thus, cloning a DNA of interest into pEMBL8 (+) and (−), followed by superinfection by phage F1, allows sequencing of both strands of the DNA of interest. Because they are smaller (4 kb) than M13 vectors, the pEMBL plasmids can accept longer inserts and provide a higher stability of the cloned DNA of interest.

A different single-stranded : double-stranded bifunctional family of plasmids has been independently constructed by Zagursky and Berman (1984). They took the replication system of M13 phage and inserted it into pBR322. Plasmid pZ150 has a 514-base pair *Rsa* fragment from M13 inserted into the *Aha*III sites of pBR322 (positions 3232 and 3251). Plasmid pZ152 has a 455-base pair *Rsa*I–*Aha*III fragment from M13, also inserted into the *Aha*III sites of pBR322 but in the opposite orientation of pZ150. Both of these plasmids replicate like pBR322. To produce single-stranded plasmid DNA, *E. coli* male cells harboring plasmid pZ150 or pZ152 are infected with wild-type M13 phage. The single-stranded plasmid-containing particles bud from the cell, and single-stranded DNA can be extracted for sequence analysis. For example, if a foreign DNA is inserted at the *Eco*RI site in plasmid pZ150, a commercially available primer which binds near this site can be used for primer-dependent sequence analysis.

D. Progressive Shortening of DNA *in Vivo*

Method 12: Progressive shortening of DNA by transposon-promoted deletions *in vivo*.

All the methods described so far for the progressive shortening of DNA involve enzymatic methods *in vitro*. Ahmed (1984, and personal communication) devised a clever scheme based on the use of transposon-promoted deletions *in vivo*. The method starts with a DNA fragment of interest cloned in a plasmid (pAA3.7X), which includes the essential sequences from transposon Tn9 and a segment of the *gal* operon (carrying the *K* and *T* genes). As seen in Fig. 10, the DNA to be sequenced is inserted in the *tet* gene (in the *Bam*HI site in this example), thus inactivating the *tet* gene. Transposon-promoted *in vivo* deletions start from a fixed point within the IS1-L sequence and end at variable sites on the adjoining DNA. Thus, the original cloned DNA is divided into many overlapping segments. If a transposon with a terminus X (located at the right-hand side of IS1-L) is located next to a DNA segment of the sequence ABCD, these deletions would generate overlapping segments of the type X-BCD (e.g., in d1 clone), X-CD, and X-D. Thus, using a primer complementary to a portion of the IS1-L sequence, it is possible to determine the nucleotide sequences of B, C, and D in different clones, and assemble them into the complete sequence.

The deletions are isolated by positive selection of plasmid pAA3.7X which carries the *K* and *T* genes of the *gal* operon (Fig. 10). Growth of cells containing this plasmid in the presence of galactose results in the accumulation of UDP-galactose, which kills the cells. Thus, plating these cells on galactose-containing media selects for only those cells in which the activity of the *gal* genes is deleted. In this way, many galactose-resistant (*Gal*[R]) clones, which have deleted the *K* and *T* genes and ended randomly in the cloned fragment of interest, can be positively selected.

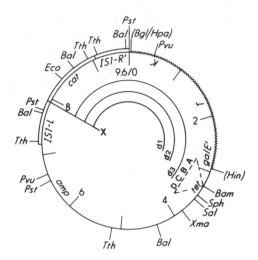

Fig. 10. Structure of the sequencing vector pAA3.7X (Ahmed, 1984). The plasmid vector pAA3.7X carries a portion of the transposon Tn9, between IS1-L and IS1-R. The plasmid without an insert has a phenotype of Amp^R, Cam^R, Gal^S, and Tet^R. After inserting the DNA to be sequenced (ABCD, dotted triangle) at the BamHI site, the plasmid has a phenotype of Amp^R, Cam^R, Gal^S and Tet^S. Transposon-promoted deletions start at the right-hand border of IS1-L (at position 7.8) and extend clockwise, deleting the cat and then K and T genes before entering the insert DNA. The phenotype of these plasmids is Amp^R, Cam^S, Gal^R, and Tet^S. In progressively shortened plasmid d1, region A of the insert DNA is deleted, thus terminus X is adjacent to BCD. Similarly, in progressively shortened plasmid d2, X is adjacent to region CD.

From these clones, deletions with overlapping end points are isolated by fractionating mixed plasmid preparations by electrophoresis in low-melting-point agarose gels. Plasmids containing deletions of progressively increasing size are recovered from successive gel slices after transformation.

For sequence analysis, the plasmid preparations are purified by the alkali method of Marko *et al.* (1982) followed by chromatography on NACS PREPAC columns (Bethesda Research Laboratories). The plasmid DNA is denatured by alkali, annealed to a IS1 primer, and sequenced directly by the dideoxynucleotide chain-termination method. The complete sequence of the cloned DNA of interest (insert DNA) is assembled from overlaps that occur between the sequences of adjacent deletions.

In the method of Ahmed (1984), there may be problems during sequencing which are common in sequencing any double-stranded DNA. Namely, during annealing of the primer to the denatured plasmid DNA, the two strands of the plasmid may tend to reanneal and exclude the primer from binding efficiently. We believe that one way to overcome this potential problem is to clone the F1 phage (Dente *et al.*, 1983) or M13 phage (Zagursky and Berman, 1984) origin of replication into the pAA3.7X plasmid. The modified plasmid (designated pAA-PZ1) may then be used as a vector to clone the DNA of interest. After *in vivo* deletion of the cloned gene in pAA-PZ1, one can superinfect these plasmids with F1 or M13 phage. The resulting single-stranded DNA can be more easily sequenced by the primer-dependent dideoxynucleotide chain-termination method (Peng and Wu, 1986).

V. DISCUSSION

A. Requirements for Successful DNA Sequence Analysis

Methods for DNA sequence analysis involve nucleic acid enzymology (i.e., restriction endonucleases and other enzymes for DNA manipulation), nucleic acid chemistry (i.e., nucleotide synthesis, modification, and degradation), gel electrophoresis technology (i.e., agarose gels and polyacrylamide gels), and vector construction (i.e., plasmids and single-stranded phage). Success in any one of the methods reviewed in this article depends on several factors. The common factors include (1) high-quality enzymes (free of contaminant nucleases) and pure isotope-labeled materials (with little or no breakdown products or inhibitors), (2) homogeneous and intact DNA (free of contaminating DNA and free of nicks or breaks), and (3) smooth sequencing gels (no bubbles or uneven heating potential) and good techniques in loading the samples onto the gel. Any investigator who wishes to use any of these rapid methods must read the original article and the related papers.

1. Enzymes and Radioactive Nucleotides

To date, a large number of useful enzymes and isotope-labeled nucleotides are available from the commercial suppliers. These products are usually of high quality. Occasionally a poor lot does appear, however, and the investigator must be aware of this possibility. Ideally, each investigator should devise simple test systems, using known samples which should work as positive controls before applying a new lot of commercial product on an unknown DNA sample.

In the original dideoxynucleotide method of DNA sequencing (Sanger *et al.*, 1977), ^{32}P-labeled dNTPs are used. In an improved procedure, ^{35}S-labeled deoxyadenosine 5′-[α-thio]triphosphate (^{35}S-dATPαS) is used (Biggin *et al.*, 1983) in place of [α-^{32}P]dATP. Use of the ^{35}S nucleotide in sequence analysis has the following advantages. (a) Autoradiographs of ^{35}S-labeled DNA give sharper bands because of the short path length of the β-particles emitted by ^{35}S. Thus, the larger DNA fragments are better resolved on autoradiographs and more sequence information can be obtained from the gel. (b) The longer half-life of ^{35}S relative to ^{32}P (87 days versus 14 days), allows [^{35}S]dATP and ^{35}S-labeled DNA to be stored for a longer period of time. (c) The lower energy from ^{35}S reduces the worker's radiation exposure during the sequencing procedure.

2. DNA Isolation

For large-scale isolation of double-stranded DNA, the CsCl–ethidium bromide centrifugation method to separate the supercoiled DNA from the contaminating RNA or the chromosomal DNA is recommended. The purificiation of restriction DNA fragments on a low melting-point agarose gel is generally better than using the ordinary agarose gel, since the DNA is relatively free of agaropectin-related and other inhibitors. For small-scale isolation of double- or single-stranded DNA, several rapid methods are available (see procedures used or referred to in Methods 3–12).

3. Sequencing Gels

Preparation of thin polyacrylamide sequencing gels using the shark-tooth comb (perfected by Robert Yang in 1978, and quoted in Wu *et al.*, 1984) has proved extremely practical and efficient (both bubble and trouble free). To assure even heating of the sequencing gel during electrophoresis of the labeled DNA molecules, a 3-mm-thick aluminum sheeting clamped to either side of the glass plates is recommended.

In general, all the methods reviewed in this chapter (chemical or enzymatic) are relatively rapid and reliable. Most of these methods are intended for determining long sequences in a short period of time by either the random shotgun or the nonrandom systematic subcloning-sequencing strategies. However, all of them rely strongly on the limited resolving power of polyacrylamide gel electrophoresis. Beck and Pohl (1984) have recently developed a method which may improve the legibility of the sequencing data by transferring the DNA molecules onto an immobilizing matrix during electrophoresis. In this method, a durable blotting membrane, such as NA45 or Biodyne A, moves with constant speed across the end of a short denaturing gel and collects the DNA molecules as they reach the end of the gel. The direct blotting electrophoresis apparatus is shown schematically in Fig. 11. The major difference from the usual gel electrophoresis device is a motor-driven conveyor belt across the lower end of the gel. The conveyor belt, made of a fine-meshed nylon net, serves as a movable support for the blotting membrane. Thereby, the temporal order of transfer of DNA from the gel is converted to a geometric one on the blotting membrane.

With a standard gel, several loadings of samples are usually needed to

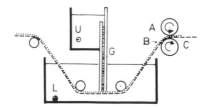

Fig. 11. Schematic drawing of a direct blotting electrophoresis device for sequencing DNA (Beck and Pohl, 1984). A motor-driven axis (A) moves the conveyor belt (C), with a blotting membrane (B) attached to it, at constant speed across the end of a short separating gel (G). The lower (L) and upper (U) buffer chambers contain the electrodes, which are connected to a power supply.

exploit the resolving power of the gel; up to 600 nucleotides can be analyzed on a 80-cm gel. With the direct blotting method, only a single loading is required to resolve the sequence of over 1000 nucleotides. The drastic shortening of the gel length (6–20 cm instead of 40–80 cm) considerably facilitates the preparation of gels, including the exclusion of air bubbles. Correspondingly, much lower voltage is necessary so that safety of handling is better ensured. This method can be readily developed into an automatic procedure.

B. Speed in DNA Sequence Analysis

DNA sequence analysis is usually one of the last steps of a project on the isolation and analysis of a gene. Some of the projects start with the purification of the DNA; it may take months to work out the optimal conditions to obtain the cells or tissues and to purify the DNA to at least 95% purity. Other projects start with the cloning of a gene, which may take several months to a year to complete. After the initial cloning, it is often necessary to trim away unrelated DNA and to transfer the gene from the original λDNA vector to a smaller vector, such as a plasmid or a M13 phage for sequence analysis.

There have been inconsistencies in reporting the speed of DNA sequencing. When the ability to sequence 2000 nucleotides a day is claimed, it usually means that the sequencing procedure, starting with a set of pure DNAs (plasmids or fragments), can be completed in less than 24 hr. Using a 40-cm-long gel with two loadings of each sample, one can usually read 250 nucleotides. If there are eight samples, 2000 nucleotides could be read. However, if one includes the time for exposing the radioautogram and several hours needed to read and record the sequence, it really takes 2 days to sequence 2000 nucleotides. The length of time required to sequence a fragment of DNA of interest is much longer than that needed to carry out the sequencing reactions and run the gel. One should really include the time required for the manipulation of the DNA of interest prior to the sequencing reactions. If the DNA of interest is already cloned in a vector and the length of DNA to be sequenced is longer than 1000 base pairs, the manipulation of the DNA usually requires digestion and subcloning. The time required for subcloning and identifying the subclones, and for putting together the sequences of the subclones, must be included in the time required for DNA sequencing.

What is usually not mentioned is that of the 2000 nucleotides, a sequence of 20 or more nucleotides is likely to be ambiguous. Repeating the sequencing procedure once or several times will be necessary to get close to 100% of the sequence confirmed on one strand. Another week may be necessary to determine the sequence of the complementary strand to obtain truly reliable sequence information. The above estimate is an optimistic one, because in reality very few investigators can obtain publishable results of a 2000-base-pair-long DNA in 2 weeks.

One may ask which one of the methods described in this chapter is easiest to use and fastest in giving the sequencing results. The answer depends on the protocol of the method and on the size of the DNA to be sequenced. With the availability of chemically synthesized primers, the primer-extension approach would seem to be the easiest for DNA sequence analysis. One possible problem with this approach is that secondary structures with long stems may be formed at certain regions of the template DNA, which may make it difficult for the DNA polymerase to incorporate nucleotides beyond these regions. For sequencing these regions, the chemical method may be needed.

We believe that the best strategy for sequencing a long DNA (between 2000 and 5000 base pairs) is to use one of the primer-extension methods to obtain the majority of the sequence, which would also serve to provide a physical map of the DNA. Any region with a long stem-and-loop structure that resists analysis, or gives abnormal spacing of the bands on the gel, can be tackled by the chemical cleavage method using the appropriate subclone. DNA molecules much longer than 5000 base pairs can be cleaved by restriction enzymes to give shorter fragments, which can be cloned and sequenced separately.

ACKNOWLEDGMENTS

This work was supported by a research Grant GM 29179 from the National Institutes of Health, U.S. Public Health Service, and a Grant RF 84066, Allocation No. 3, from the Rockefeller Foundation.

REFERENCES

Ahmed, A. (1984). *J. Mol. Biol.* **178,** 941–948.
Anderson, S. (1981). *Nucleic Acids Res.* **9,** 3015–3027.
Barnes, W. M. (1980). *Genet. Eng.* **2,** 185–200.
Beck, S., and Pohl, F. M. (1984). *EMBO J.* **3,** 2905–2909.
Biggin, M. D., Gibson, T. J., and Hong, G. F. (1983). *Proc. Natl. Acad. Sci. U.S.A.* **80,** 3963–3965.
Dale, R. M. K., McClure, B. A., and Houchins, J. P. (1985). *Plasmid* **13,** 31–40.
Deininger, P. L. (1983). *Anal. Biochem.* **129,** 216–226.
Dente, L., Cesareni, G., and Cortese, R. (1983). *Nucleic Acids Res.* **11,** 1645–1655.
Englund, P. T. (1972). *J. Mol. Biol.* **66,** 209–224.
Frischauf, A. M., Garoff, H., and Lehrach, H. (1980). *Nucleic Acids Res.* **8,** 5541–5549.
Gronenborn, B., and Messing, J. (1978). *Nature (London)* **272,** 375–377.
Guo, L.-H., and Wu, R. (1982). *Nucleic Acids Res.* **10,** 2065–2084.
Guo, L.-H., and Wu, R. (1983). *In* "Methods in Enzymology" (R. Wu, L. Grossman, and K. Moldave, eds.), Vol. 100, pp. 60–96. Academic Press, New York.
Guo, L.-H., Yang, R. C. A., and Wu, R. (1983). *Nucleic Acids Res.* **11,** 5521–5540.
Hartley, J., Chen, K. K., and Donelson, J. E. (1982). *Nucleic Acids Res.* **10,** 4009–4025.
Henikoff, C. (1984). *Gene* **28,** 351–359.
Hindley, J. (1983). "DNA Sequencing." Elsevier Biomedical Press, Amsterdam.
Hong, G. F. (1982). *J. Mol. Biol.* **158,** 539–549.
Lin, H. C., Lei, S. P., and Wilcox, G. (1985). *Anal. Biochem.* **147,** 114–119.
Marko, M. A., Chipperfield, R., and Birnboim, H. C. (1982). *Anal. Biochem.* **121,** 382–387.
Maxam, A. M., and Gilbert, W. (1977). *Proc. Natl. Acad. Sci. U.S.A.* **74,** 560–564.
Maxam, A. M., and Gilbert, W. (1980). *In* "Methods in Enzymology" (L. Grossman and K. Moldave, eds.), Vol. 65, pp. 499–560. Academic Press, New York.
Messing, J., Crea, R., and Seeburg, P. H. (1981). *Nucleic Acids Res.* **9,** 309–321.
Murray, K., and Murray, N. E. (1973). *Nature (London), New Biol.* **243,** 134–139.
Narang, S. A., Brousseau, R., Hsiung, H. M., Hess, B., and Wu, R. (1980). *Nucleic Acids Symp. Ser.* **7,** 377–385.
Peng, Z.-G., and Wu, R. (1986). *Gene* **45,** 247–252.
Poncz, M., Solowiejczyk, D., Ballentine, M., Schwartz, F., and Surrey, S. (1982). *Proc. Natl. Acad. Sci. U.S.A.* **79,** 4298–4302.
Richardson, C. C., Lehman, I. R., and Kornberg, A. (1964). *J. Biol. Chem.* **239,** 251–258.
Rüther, U., Koenen, M., Otto, K., and Muller-Hill, B. (1981). *Nucleic Acids, Res.* **9,** 4087–4098.
Sanger, F., and Coulson, A. R. (1975). *J. Mol. Biol.* **94,** 441–448.
Sanger, F., and Coulson, A. R. (1978). *FEBS Lett.* **87,** 107–110.
Sanger, F., Donelson, J. E., Coulson, A. R., Kossel, H., and Fischer, D. (1973). *Proc. Natl. Acad. Sci. U.S.A.* **70,** 1209–1213.
Sanger, F., Nicklen, S., and Coulson, A. R. (1977). *Proc. Natl. Acad. Sci. U.S.A.* **74,** 5463–5467.
Sanger, F., Coulson, A. R., Barrell, B. G., Smith, A. J. H., and Roe, B. A. (1980). *J. Mol. Biol.* **143,** 161–178.
Smith, A. J. R. (1979). *Nucleic Acids Res.* **6,** 831–848.
Smith, D. R., and Calvo, J. M. (1980). *Nucleic Acids Res.* **8,** 2255–2273.
Smith, M., Leung, D. W., Gillam, S., Astell, C. R., Montgomery, D. L., and Hall, B. D. (1979). *Cell (Cambridge, Mass.)* **16,** 753–761.
Wallace, R. B., Johnson, M. J., Suggs, S. V., Miyoshi, K., Bhatt, R., and Itakura, K. (1981). *Gene* **16,** 21–26.

Wu, R. (1972). *Nature (London), New Biol.* **236,** 198–200.

Wu, R. (1978). *Annu. Rev. Biochem.* **47,** 607–634.

Wu, R., and Kaiser, A. D. (1968). *J. Mol. Biol.* **35,** 523–537.

Wu, R., and Taylor, E. (1971). *J. Mol. Biol.* **57,** 491–511.

Wu, R., Tu, C. D., and Padmanabhan, P. (1973). *Biochem. Biophys. Res. Commun.* **55,** 1092–1099.

Wu, R., Guo, L.-H., Georges, F., and Narang, S. A. (1983). *Trans. N. Y. Acad. Sci.* **41,** 253–267.

Wu, R., Wu, N.-H., Zahre, H., Georges, F., and Narang, S. (1984). *In* "Oligonucleotide Synthesis: A Practical Approach" (M. J. Gait, ed.), pp. 135–151. IRL Press, Washington, D.C.

Yang, R. C. A., and Wu, R. (1979). *Virology* **92,** 340–352.

Yang, R. C. A., Guo, L.-H., and Wu, R. (1983). *In* "Frontiers in Biochemical and Biophysical Studies of Proteins and Membranes" (T. Y. Liu, ed.), pp. 5–21. Elsevier/North-Holland, New York.

Yanisch-Perron, C., Vieira, J., and Messing, J. (1985). *Gene* **33,** 103–119.

Zagursky, R., and Berman, M. (1984). *Gene* **27,** 183–191.

Zain, B. S., and Roberts, R. J. (1979). *J. Mol. Biol.* **131,** 341–352.

7

Direct, Rapid RNA
Sequence Analysis

DEBRA ANN PEATTIE
Department of Tropical Public Health
Harvard School of Public Health
Boston, Massachusetts 02115

I. INTRODUCTION

The original methods of RNA sequence analysis were based on enzymatic production and chromatographic separation of overlapping oligonucleotide fragments from within an RNA molecule followed by identification of the mononucleotides comprising the oligomer (for reviews, see Barrell, 1971; Brownlee, 1972). Over the past decade the field of nucleic acid sequencing has changed dramatically, however, and RNA molecules now can be sequenced in a variety of more streamlined fashions. Most of the more recent advances in RNA sequencing have involved one-dimensional electrophoretic separation of ^{32}P-end-labeled oligoribonucleotides on polyacrylamide gels (e.g., Brownlee and Cartwright, 1977; Donis-Keller *et al.*, 1977; Simoncits *et al.*, 1977; Lockhard *et al.*, 1978; Ross and Brimacombe, 1978; McGeogh and Turnbull, 1978; Zimmern and Kaesberg, 1978; Kramer and Mills, 1978; Peattie, 1979).

In this chapter I will discuss two of these methods for determining the nucleotide sequences of RNA molecules rapidly: the chemical method (Peattie, 1979) and the enzymatic method (Donis-Keller et al., 1977; Simoncits *et al.*, 1977; Ross and Brimacombe, 1978; Krupp and Gross, 1979). Both methods are direct and degradative, i.e., they rely on frag-

Synthesis and Applications
of DNA and RNA

mentation of the RNA substrate itself. The other sequencing methods rely on synthesizing DNA (Brownlee and Cartwright, 1977; McHeogh and Turnbull, 1978; Zimmern and Kaesberg, 1978) or RNA (Kramer and Mills, 1978) from the original RNA molecule.

II. PREPARATION OF THE RNA SUBSTRATE

A. Isolation of the RNA

Because of their hydrophobicity, both ribonucleic acids and deoxyribonucleic acids are separated easily from other macromolecular cellular constituents. Several phenol extractions of a cleared cell lysate, for example, provide a several fold purification of the nucleic acids from within those cells. Ethanol precipitation in the presence of salt rapidly separates DNA from RNA on a preparative scale: Both nucleic acids will come out of solution but only the DNA, due to its much larger molecular weight, will precipitate as a fibrous, gelatinous material that can be removed from the solution with a hooked glass rod. Also, the single-strandedness of RNA is useful for separating it from DNA in a mixed solution of nucleic acids by hydroxylapatite chromatography (Bernardi, 1971).

There are many methods for separating different types of RNA molecules from each other. Quite unlike DNA molecules, RNAs can be char-acterized and classified discretely from each other by properties such as function, size, shape, modified base content, and poly(A) tails. These characteristics often serve as the basis for physical separation and purification of different RNA molecules from each other (see Peattie, 1983b).

For the purpose of isolating small ($\leq 1,000$ nucleotides) RNA molecules for direct sequencing, bulk separation on polyacrylamide gels is recommended followed by elution (Garel *et al.*, 1977; Maniatis and Efstradiatis, 1980). Methylene blue or ethidium bromide can be used to visualize the RNA within such preparative gels; alternatively, an intercalating agent can be avoided by placing the gel over a cellulose sheet impregnated with a fluorescent indicator and illuminated with an ultraviolet light (Hassur and Whitlock, 1974).

B. Radioactive Labeling of the RNA

In theory, either the 3' or 5' terminus of the purified RNA can be labeled with ^{32}P for sequencing. In practice, however, certain considerations should be made.

The three-dimensional conformation of an RNA molecule can affect the accessibility of the ends and, consequently, the ease with which they can be labeled with a radioactive moiety. For example, the structure of the amino acid acceptor stem makes it difficult to dephosphorylate and rephosphorylate the 5' end of tRNA molecules but very easy to transfer label to the 3' end. Conversely, the 5' end of yeast 5.8 S ribosomal RNA is dephosphorylated and rephosphorylated quite easily while the 3' end is tucked beneath an overhang (see Rubin, 1973, for proposed secondary structure). In general, if one end of the molecule proves particularly difficult to label, I suggest trying to label the opposite end.

One should also consider how far in and proximal to which end sequence determination is to be made. If the RNA molecule is large and sequence information is desired from only one end, then that end should be labeled if possible. Large RNA molecules can be fragmented into smaller pieces (Stepanova *et al.*, 1979; Donis-Keller, 1979) prior to or subsequent to radioactive labeling if desired.

The chemically generated pyrimidine cleavage patterns of an RNA molecule labeled at the 5' end are not as clean as those generated from a 3' end-labeled molecule (Peattie, 1979); however, this does not interfere with unambiguous sequence determination. This phenomenon is manifested as a light background of nonspecific strand scissions superimposed upon the uridine pattern while the cytidine pattern has the same background of random strand scission without specific fragmentation at cytidines. Because cleavages at guanosines, adenosines, and uridines are unambiguous, the cytidines are determined by the bands within the random ladder. This phenomenon is not understood but is attributed to a reaction between hydrazine and the 2'-hydroxyl group of the ribose ring which remains attached to the 5' half of the cleaved molecule (see sections III,A and III, B and Peattie, 1983b). The site of labeling does not affect the pyrimidine lanes of sequencing patterns generated enzymatically.

1. 3' End-Labeling

T4 RNA ligase and [5'-^{32}P]pCp can be used to label many RNA molecules at their 3' termini very efficiently (England and Uhlenbeck, 1978a,b). Details of this procedure have been described elsewhere (Peattie, 1979, 1983a). tRNA molecules also can be labeled very efficiently at their 3' ends with [α-^{32}P]ATP and *Escherichia coli* nucleotidyltransferase (Carre *et al.*, 1970). This enzyme fills out the 3' terminal CCA moiety of *E. coli* tRNA molecules and also uses many non-*E. coli* tRNAs as substrates. Purified tRNAs often can be end-filled directly due to incomplete CCA ends (Carre *et al.*, 1970; Peattie and Gilbert, 1980), and snake venom phosphodiesterase can be used to generate incomplete ends for any tRNA molecule (Silberklang *et al.*, 1977).

2.   5' End-Labeling

The 5' terminal phosphate must be removed prior to replacing it with a radioactive phosphate moiety. Usually this is done with calf intestinal phosphatase or with bacterial alkaline phosphatase, and conditions for dephosphorylation and subsequent phosphorylation with [γ-^{32}P]ATP have been described by Simsek et al. (1973), Silberklang et al. (1979), and Chaconas and van de Sande (1980). If the 5' terminus of the molecule is difficult to dephosphorylate due to higher-order structure of the RNA, it is particularly important to use bacterial alkaline phosphatase. Unlike the calf intestinal enzyme, the bacterial phosphatase can be used at high temperatures (up to 68°C) which will help destabilize the structure of the RNA and will likely make the 5' end more accessible.

Most eukaryotic and viral messenger RNA molecules possess a 5' cap structure [m^{7}G(5')ppp(5')X; Shatkin, 1976] that must be removed prior to labeling with ^{32}P. These structures can be removed chemically using sodium periodate to oxidize the 5' terminus and aniline to remove the damaged cap (Muthukrishnan and Shatkin, 1975; Flavell et al., 1980; Cowie et al., 1981), or they can be removed enzymatically using tobacco acid pyrophosphatase to hydrolyze the m^{7}G(5')ppp moiety (Efstradiatis et al., 1977). The decapped molecules then can be labeled using [γ-^{32}P]ATP and T4 polynucleotide kinase following phosphatase removal of the remaining 5'-phosphate (Flavell et al., 1980).

III.   CHEMICAL SEQUENCING OF RNA MOLECULES

In this method, RNA molecules end-labeled with ^{32}P are damaged specifically in a limited fashion using base-specific chemical reagents. The damaged molecules are then fragmented chemically at the sites of modification. The resultant products constitute a nested set of radioactive RNA fragments and are resolved by length on a thin polyacrylamide gel (Sanger and Coulson, 1978) and visualized by autoradiography. Details of the experimental procedures are described in the appendix to this chapter and in Peattie (1979, 1983a). The probable chemistry of the reactions involved are considered here as well as in Peattie (1983b).

The strand cleavages are generated by two separate chemical reactions. First, the specific bases are damaged in a random and limited fashion; second, the RNA backbone is broken at the sites of damage. Dimethyl sulfate modifies guanosines, diethyl pyrocarbonate modifies both adenosines and guanosines, and hydrazine attacks both uridines and cytidines. The absence or presence of sodium chloride determines the base specificity of the hydrazine attack. After each of the initial chemical modifications, strand-scission is induced in an acidic aniline solution.

A. Base-Specific Chemical Modification Reactions

1. Guanosine

Dimethyl sulfate alkylates guanosine preferentially at the N-7 position (Brookes and Lawley, 1961; Lawley and Brookes, 1963) and perturbs the electron resonance of the purine ring (Lawley and Brookes, 1963; Kochetov and Budovskii, 1972). This destabilized ring reduces easily in a dilute sodium borohydride solution (Pochon *et al.*, 1970; Wintermeyer and Zachau, 1975), and the resultant m^7-dihydroguanosine hydrolyzes partially or completely to unknown products (Fig. 1). This site is then susceptible to aniline-induced strand scission (Wintermeyer and Zachau, 1975).

2. Adenosine

Diethyl pyrocarbonate carbethoxylates the base nitrogens of nucleic acids, particularly the N-7 atoms of adenosines and guanosines (Leonard *et al.*, 1971; Vincze *et al.*, 1973; Ehrenberg *et al.*, 1976). This modification

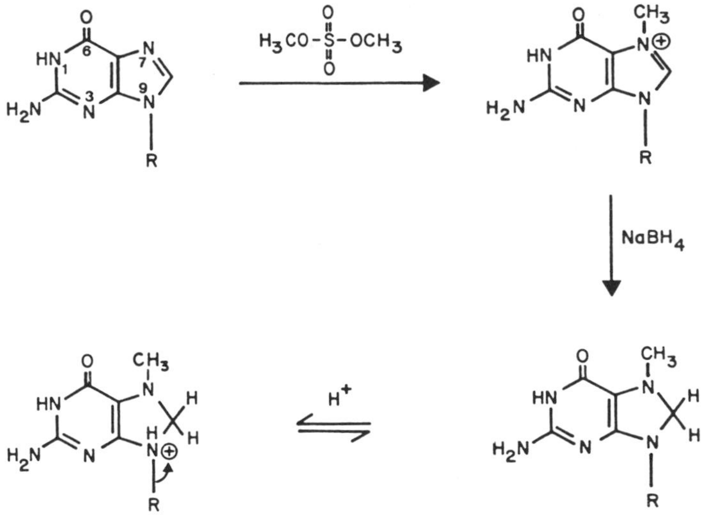

Fig. 1. Dimethyl sulfate alkylates guanosine preferentially at the N-7 position, perturbing the electron resonance of the purine ring. Sodium borohydride reduces this product, and the resultant 7-methyldihydroguanosine is susceptible to attack in an acidic aniline solution (see Fig. 5). R = β-D-Ribofuranosyl.

destroys the resonance of the heterocyclic ring systems, and the imidazole ring opens between atoms N-7 and C-8 (Leonard *et al.*, 1971) (see Fig. 2). This creates a site for aniline-induced strand scission.

3. Uridine

Unprotonated hydrazine opens pyrimidine rings by nucleophilic addition at the 5,6-double bond (Brown, 1974; Cashmore and Peterson, 1978). A pyrazolone ring product is released into solution (Levene and Bass, 1927; Baron and Brown, 1955; Cashmore and Peterson, 1978) (see Fig. 3), and the ribose ring is left exposed and susceptible to aniline attack.

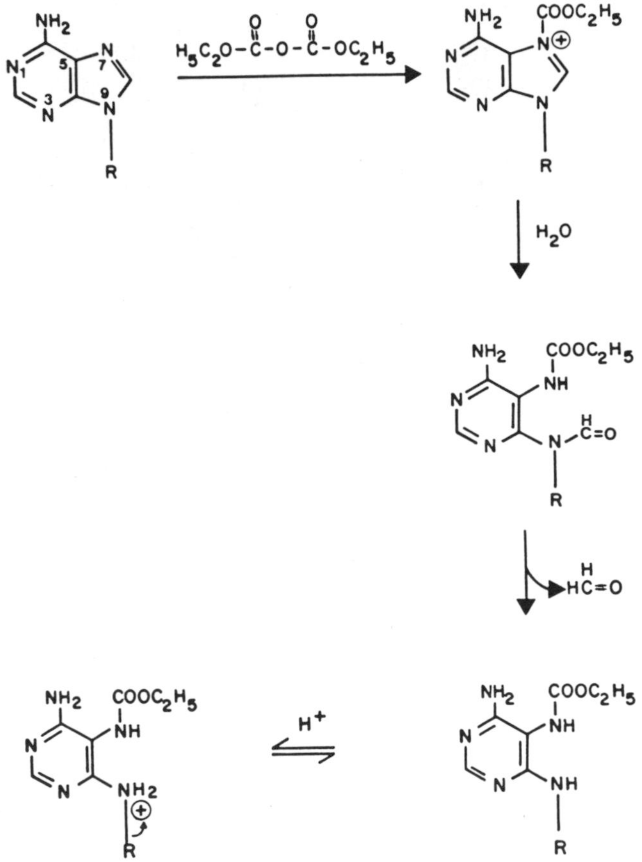

Fig. 2. Diethyl pyrocarbonate carbethoxylates the N-7 atom of adenosine (and, to a lesser degree, the N-7 atom of guanosine). The imidiazole ring opens, formaldehyde is released, and the site is susceptible to aniline-induced strand scission (see Fig. 5). R = β-D-Ribofuranosyl.

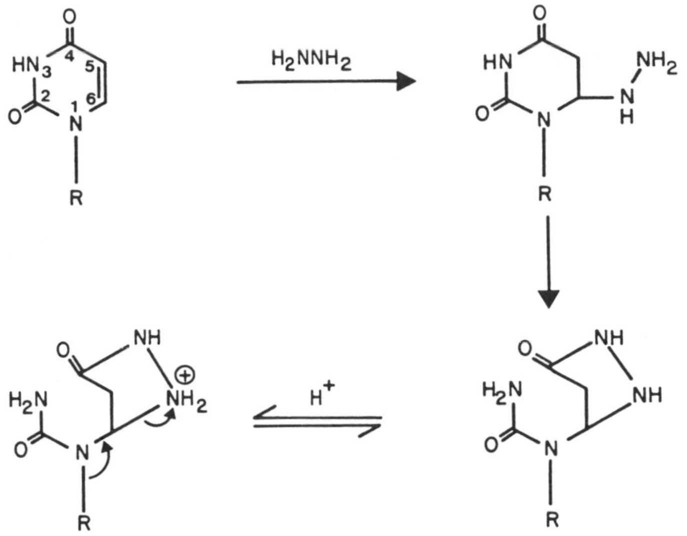

Fig. 3. Hydrazine attacks the C-4 and C-6 atoms of uridine, releasing a pyrazolone ring product into solution and exposing a site for aniline-induced strand scission (see Fig. 5). R = β-D-Ribofuranosyl.

4. Cytidine

(a) Three molar sodium chloride in anhydrous hydrazine results in significant hydrazinolysis of cytidines (Peattie, 1979). The salt-induced suppression of uridine hydrazinolysis is not understood but is analogous to the decreased hydrazinolysis of thymidines in the presence of sodium chloride (Maxam and Gilbert, 1977, 1980). Following hydrazinolysis (Fig. 4), cytidines are susceptible to aniline-induced strand scission.

(b) Dimethyl sulfate alkylates the N-3 atom of cytidines (Lawley and Brookes, 1963) as well as the N-7 atom of guanosines. This positive charge on the pyrimidine ring makes m^3C susceptible to rapid hydrazinolysis in aqueous hydrazine (Fig. 4) and, subsequently, to aniline-induced strand scission (Peattie and Gilbert, 1980). The methylated guanosines are not perturbed by the hydrazine and aniline reactions, but strand scission will occur at uridines during these steps.

B. Chemical Strand-Scission Reaction

Primary amines induce strand scission in depurinated or depyrimidinated nucleic acids (Philippsen et al., 1968; Vanyushin and Bur'yanov, 1969; Tuchinskii et al., 1970). This reaction occurs via Schiff base formation with the open-ring tautomer of the ribose moiety with subsequent β-

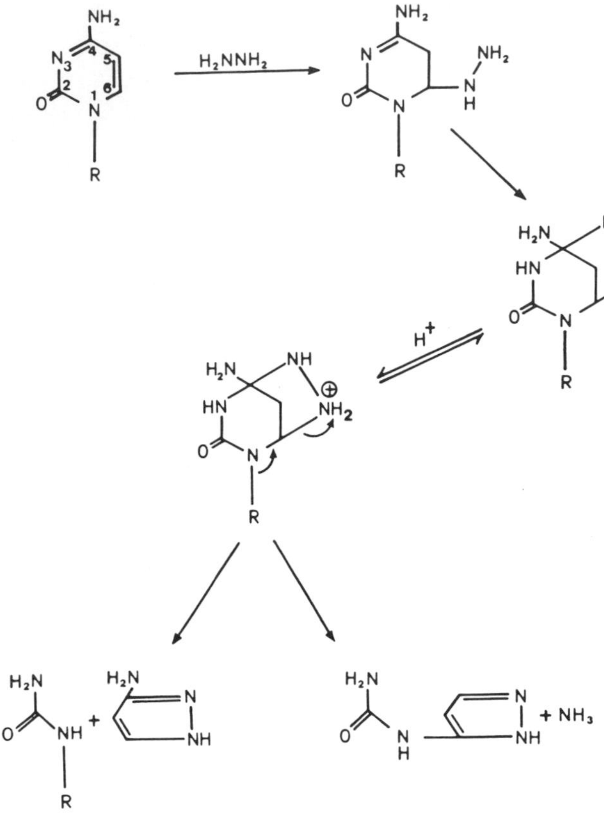

Fig. 4. Hydrazine attacks the C-4 and C-6 atoms of cytidine and releases 3-aminopyrazole (lower left) and 3-ureidopyrazole (lower right) into solution. The site is then susceptible to aniline-induced strand scission (see Fig. 5). R = β-D-Ribofuranosyl.

elimination and cleavage of the phosphodiester bond (Whitfeld, 1954, 1965; Burton and Peterson, 1960; Yu and Zamecnik, 1960; Khym, 1963; Neu and Heppel, 1964; Steinschneider and Fraenkel-Conrat, 1966). An aniline acetate solution at pH 4.5 effects this reaction efficiently (Fig. 5) without random hydrolysis and provides a volatile medium for lyophilization (Peattie, 1979).

C. Procedures and Considerations

As stated in the introduction to this chapter, all procedures for the reactions described above are in the appendix to this chapter and in

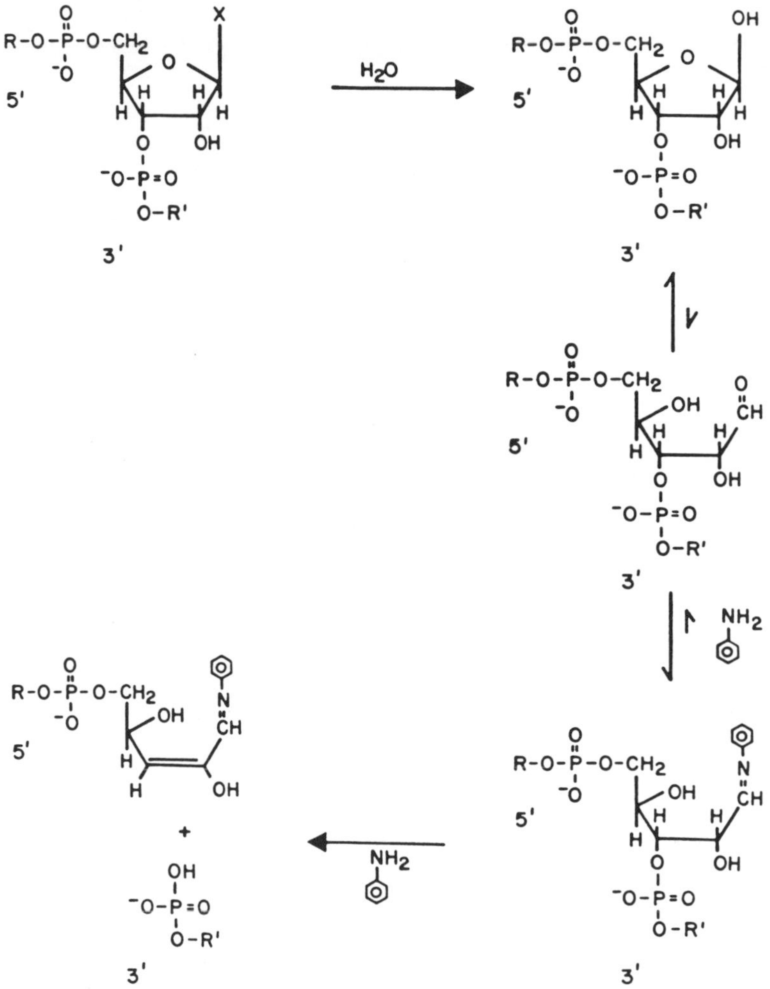

Fig. 5. Once the glycosidic bond between the chemically modified base (X) and the ribose ring is hydrolyzed, the exposed ribose moiety is free to form its open-ring tautomer. Aniline reacts with the free aldehyde group of this tautomer to form an aldimine (a Schiff base), and the 3' phosphodiester bond is cleaved via β-elimination. R and R' are the respective 5' and 3' portions of the RNA molecule proximal to the exposed ribose residue.

Peattie (1979). Diethyl pyrocarbonate, dimethyl sulfate, and hydrazine are dangerous chemicals and should be handled with extreme caution in the hood.

The chemical modification reactions can be used on impure preparations or mixed populations of RNA molecules as well as on purified

RNAs. Because the base-specific modification reactions do not induce strand scission, they can be done prior to purifying the RNA for sequencing (Peattie and Herr, 1981; Peattie, 1983b) as well as prior to the radioactive end-labeling. For all general purposes, the chemically modified RNA molecules are as stable as unmodified RNAs. The RNA molecules can be renatured after the limited modification, and this property is useful for locating and identifying specific nucleotides crucial to sites of interaction and/or contact within RNA–protein complexes (Peattie and Herr, 1981; Peattie *et al.*, 1981; Peattie, 1983a).

The absence of a band or the ubiquitous presence of a band in the lanes of a chemically generated sequence pattern indicates that a modified ribonucleotide is present. Unambiguous identification of the modified nucleotide involves more extensive characterization, usually involving two-dimensional chromatography (Hall, 1971; Nishimura, 1972). RNA sequenced as cDNA provides no information about modified nucleotides within the original RNA molecule.

IV. ENZYMATIC SEQUENCING OF RNA MOLECULES

Limited digestion of ^{32}P-end-labeled RNAs with base-specific ribonucleases also provides a means to sequence RNA molecules directly. Several reports of this approach have been made (Donis-Keller *et al.*, 1977; Simoncits *et al.*, 1977; Ross and Brimacombe, 1978; Krupp and Gross, 1979), and each involves separating the RNA fragments in one dimension by polyacrylamide gel electrophoresis.

Ribonuclease T_1 cleaves the end-labeled molecules at guanosines, ribonuclease U_2 cleaves at adenosines, and pancreatic RNase or an extracellular RNase from *Bacillus cereus* cleaves both uridines and cytidines. Nucleases from *Neurospora crassa* and *Staphylococcus aureus* also can be used to identify pyrimidines and to distinguish between cytidines and uridines (Krupp and Gross, 1979); a chicken liver ribonuclease can be used for this purpose as well (Levy and Karpetsky, 1980; Boguski *et al.*, 1980). Many companies that sell reagents for molecular biology, e.g., Bethesda Research Laboratories, Boehringer-Mannheim Biochemicals, and Pharmacia P-L, sell self-explanatory kits containing all of the ribonucleases used for sequencing RNA enzymatically.

In each of the enzymatic sequencing methods, the RNA samples are loaded directly onto a polyacrylamide sequencing gel immediately following enzymatic digestion. Because the enzymatic digestion is a one-step procedure, the enzymatic sequencing reactions are far faster and simpler than the chemical ones. Higher-order interactions within the RNA impose the major limitation to enzymatic sequencing, and in practice most RNA molecules cannot be sequenced entirely by enzymatic methods. Given the pH and temperature constraints for required enzymatic function, regions of very stable secondary structure within the RNA can remain intact during digestion and can prevent nucleolytic digestion entirely. This results in gaps in the sequence display which can be filled in only by using the chemical sequencing reactions.

V. PROBING RNA CONFORMATION

A. Adjusting the Chemical Sequencing Reactions for Probing Higher-Order Structure

The chemical sequencing reactions for guanosines, adenosines, and cytidines can be adapted to probe the higher-order interactions within an RNA molecule (Peattie and Gilbert, 1980; Christensen et al., 1985) or within an RNA–protein complex (Peattie and Herr, 1981; Peattie et al., 1981). This is described in detail in Peattie (1983a). The steps are essentially the same as those used for sequencing, but the reaction buffers and temperatures are altered such that the RNA molecule remains undenatured throughout the initial chemical reaction. Unlike the sequencing reactions, the probing reactions reflect secondary interactions (base-pairing) and tertiary interactions (base-pairing, ionic interactions, base-stacking) within the molecule: Only those sites *not* involved in higher-order interactions are susceptible to modification and subsequent strand scission. Loading the sequencing reactions adjacent to the probing reactions on a polyacrylamide gel displays the sequence of the molecule and identifies regions involved in higher-order structure. These regions appear as gaps within the nucleotide sequence. Because tertiary interactions are less stable than secondary ones, a profile of progressively less stringent reaction conditions allows one to distinguish between them (de Bruijn and Klug, 1983). In addition, because the initial modifications can be done prior to isolating and end-labeling the RNA, there is no need to disrupt an RNA–protein complex before probing its structure.

1. Guanosine

The N-7 atom methylated by dimethyl sulfate (Fig. 1) is not involved in secondary Watson–Crick base pairing but is often involved in tertiary interactions. This reaction can monitor the base-stacking interactions, ion coordination interactions, and non-Watson–Crick hydrogen bonds of a native RNA molecule (Peattie and Gilbert, 1980).

2. Adenosine

As with dimethyl sulfate-induced methylation of guanosines, diethyl pyrocarbonate-induced carbethoxylation of adenosines affects primarily the N-7 atom of the purine (Fig. 2). Unlike the guanosine reaction, however, the adenosine reaction reflects both secondary and tertiary interactions by monitoring the stacking of adenosines (Peattie and Gilbert, 1980). Although diethyl pyrocarbonate modifies guanosines at the acidic pH and high temperature used for the sequencing reactions, it does not modify them significantly at the neutral pHs and lower temperatures used for probing higher-order structure.

3. Cytidine Reaction

Only the cytidine cleavage reaction utilizing dimethyl sulfate for the initial chemical modification can be used for probing conformational structure. If titrated to a neutral pH, hydrazine is protonated and will not attack the 5,6-double bond of the pyrimidines (Figs. 3 and 4); hence, hydrazinolysis of pyrimidines is impossible in less than basic solutions. Dimethyl sulfate alkylates N-3 atoms of cytidines not involved in secondary or tertiary interactions, and strand scission is induced at these methyl-3-cytidines by subsequent hydrazine and aniline reactions (Peattie and Gilbert, 1980). Uridine cleavage results during the hydrazine and aniline steps but does not reflect conformation because the high pH of the hydrazine reaction denatures the RNA.

B. Using Enzymatic Digestions to Localize Regions of Higher-Order Structure

As mentioned above, gaps within enzymatically generated sequencing pattern identify regions of stable secondary structure within an RNA molecule. In addition, the single-strand-specific ribonucleases A, T_1, T_2, and S_1 can be used to localize unstructured and accessible sequences within an RNA molecule (Douthwaite et al., 1983) while a double-strand-specific ribonuclease from Naja naja oxiana cobra venom (Vassilenko and Babkina, 1965) is useful for localizing double-helical regions within the native conformer (Douthwaite et al., 1983).

If it is small enough, the RNA should be probed in duplicate with one set of digestions performed on 3'-end-labeled material and another on 5'-end-labeled material. This helps one distinguish between regions of the RNA that are available for digestion within the native molecule and those regions that become available only after an enzyme nick induces a structural rearrangement in the molecule (Douthwaite and Garrett, 1981). If the RNA molecule is too large to make this practical, then secondary cutting should be minimized by using a range of mild digestion conditions (Douthwaite *et al.*, 1983).

Because the enzymatic fragmentation is a one-step process, the RNA must be isolated and end-labeled before enzymatic digestion. Unlike complexes probed chemically, RNA–protein complexes to be probed enzymatically must be denatured and then renatured after end-labeling the RNA. In addition, enzymes are very useful for pinpointing regions of stable higher-order structure rapidly, even though they do not provide information about particular atoms within an RNA molecule.

VI. SUMMARY

There are two primary methods for sequencing RNA molecules directly and rapidly: the chemical method and the enzymatic method. Both procedures involve base-specific fragmentation of ^{32}P-end-labeled molecules with subsequent separation of the fragments in one dimension on denaturing polyacrylamide gels, and both methods have advantages and disadvantages.

In general, the researcher with no experience in isolating and handling RNA should begin sequencing using the enzymatic method. It is faster and simpler than the chemical method, and any inherent degradation problems will become obvious sooner and with less time spent in sample manipulation. Few RNA molecules can be sequenced completely using the enzymatic reactions alone (because of internal higher-order structures that defy enzymatic sequence determination), however, and the chemical reactions probably will be required to determine portions of the sequence. For this reason, someone experienced in handling RNA might wish to use the chemical method from the outset.

When possible, I recommend sequencing an RNA molecule directly and degradatively rather than by sequencing its cDNA analog or by using a random chain-termination method. This is because (i) a cDNA sequence will not provide information about modified nucleotides in the original RNA molecule and (ii) the degradative methods provide information about the higher-order structure of the RNA molecule. Size dictates feasibility, however, and large (more than ~1000 nucleotides) RNAs should be converted into cDNA. It can then be sequenced as such (see Chapter 6 in this volume for procedures of rapid DNA sequence analysis), or 2′,3′-dideoxynucleoside triphosphate chain terminators (McGeogh and Turnbull, 1978; Zimmern and Kaesberg, 1978) can be used originally when converting the DNA into RNA.

When probing the conformation of an RNA molecule, both the enzymatic and chemical approaches should be utilized. The single-strand-specific ribonucleases (A, T_1, T_2, and S_1) provide an efficient means to locate double-helical regions rapidly, and the chemical reactions provide a means to determine the RNA sequence within these regions. In addition, the chemical reactions allow one to assign interactions to specific atoms and to distinguish secondary interactions from tertiary ones. If the RNA molecule is small enough to be sequenced directly by the enzymatic or chemical method, the probing reactions can be done easily at the same time as the sequencing reactions.

In conclusion, I wish to point out that determining both the nucleotide sequence and conformation of an RNA molecule is crucial to understanding its fundamental biological properties. Unlike DNA, where sequence reveals all, RNA molecules are more than double helices. Sequencing an RNA molecule rarely reveals its structure or function. The conformation of a single RNA molecule can be determined empirically by methods such as the enzymatic or chemical probing methods described above; the conformation of a group of RNAs (e.g., the tRNAs, 5 S RNAs, or 16 S RNAs) can be determined empirically by these means as well as statistically by comparative sequence analysis (see Noller and Woese, 1981).

REFERENCES

Baron, F., and Brown, D. M. (1955). *J. Chem. Soc.* **3**, 2855–2860.

Barrell, B. G. (1971). *In* "Procedures in Nucleic Acid Research" (G. L. Cantoni and D. R. Davies, eds.), Vol. 2, pp. 751–812. Harper & Row, New York.

Bernardi, G. (1971). *In* "Procedures in Nucleic Acid Research" (G. L. Cantoni and D. R. Davies, eds.), Vol. 2, pp. 751–812. Harper & Row, New York.

Boguski, M. S., Hieter, P. A., and Levy, C. C. (1980). *J. Biol. Chem.* **255**, 2160–2163.

Brookes, P., and Lawley, P. D. (1961). *Biochem. J.* **80**, 496–503.

Brown, D. M. (1974). *In* "Basic Principles in Nucleic Acid Chemistry" (P. O. P. Ts'o, ed.), Vol. 2, pp. 1–90. Academic Press, New York.

Brownlee, G. G. (1972). "Determination of Sequences in RNA." Am. Elsevier, New York.

Brownlee, G. G., and Cartwright, E. M. (1977). *J. Mol. Biol.* **114**, 93–117.

Bruce, A. G., and Uhlenbeck, O. C. (1978). *Nucleic Acids Res.* **5**, 3665–3677.

Burton, K., and Peterson, G. B. (1960). *Biochem. J.* **75**, 17–27.

Carre, D. D., Litvak, S., and Chapeville, F. (1970). *Biochim. Biophys. Acta* **224**, 371–381.

Cashmore, A. R., and Peterson, G. B. (1978). *Nucleic Acids Res.* **5**, 2485–2491.

Chaconas, G., and van de Sande, J. H. (1980). *In* "Methods in Enzymology" (L. Grossman and K. Moldave, eds.), Vol. 65, pp. 75–85. Academic Press, New York.

Christensen, A., Maithiesen, M., Peattie, D., and Garrett, R. A. (1985). *Biochemistry* **24**, 2284–2291.

Cowie, A., Tyndall, C., and Kamen, R. (1981). *Nucleic Acids Res.* **9**, 6305–6322.

de Bruijn, M. H. L., and Klug, A. (1983). *EMBO J.* **2**, 1309–1321.

Donis-Keller, H. (1979). *Nucleic Acids Res.* **7**, 179–192.

Donis-Keller, H., Maxam, A. M., and Gilbert, W. (1977). *Nucleic Acids Res.* **4,** 2527–2538.

Douthwaite, S., and Garrett, R. A. (1981). *Biochemistry* **20,** 7301–7307.

Douthwaite, S., Christensen, A., and Garrett, R. A. (1983). *J. Mol. Biol.* **169,** 249–279.

Efstradiatis, A., Vournakis, J. N., Donis-Keller, H., Chaconas, G., Dougall, D. K., and Kafatos, F. C. (1977). *Nucleic Acids Res.* **4,** 4165–4174.

Ehrenberg, L., Fedorcsak, I., and Solymosy, F. (1976). *Prog. Nucleic Acid Res. Mol. Biol.* **16,** 189–262.

England, T. E., and Uhlenbeck, O. C. (1978a). *Nature (London)* **275,** 5060–5061.

England, T. E., and Uhlenbeck, O. C. (1978b). *Biochemistry* **17,** 2069–2076.

Flavell, A. J., Cowie, A., Arrand, J. R., and Kamen, R. I. (1980). *J. Virol.* **22,** 346–352.

Garel, J.-P., Garber, R. L., and Siddiqui, M. A. Q. (1977). *Biochemistry* **16,** 3618–3624.

Hall, R. H. (1971). *In* "The Modified Nucleosides in Nucleic Acids," pp. 209–253. Columbia University Press, New York.

Hassur, S. M., and Whitlock, H. W., Jr. (1974). *Anal. Biochem.* **59,** 162–164.

Khym, J. X. (1963). *Biochemistry* **2,** 344–350.

Kochetov, N. K., and Budovskii, E. I. (1972). *In* "Organic Chemistry of Nucleic Acids" (eds.), pp. Plenum, New York.

Kramer, F. R., and Mills, D. R. (1978). *Proc. Natl. Acad. Sci. U.S.A.* **75,** 5334–5338.

Krupp, G., and Gross, H. J. (1979). *Nucleic Acids Res.* **6,** 3481–3490.

Lawley, P. D., and Brookes, P. (1963). *Biochem. J.* **89,** 127–138.

Leonard, N. J., McDonald, J. J., Henderson, R. E. L., and Reichmann, M. E. (1971). *Biochemistry* **10,** 3335–3342.

Levene, P. A., and Bass, L. W. (1972). *J. Biol. Chem.* **71,** 167–172.

Levy, C. C., and Karpetsky, T. P. (1980). *J. Biol. Chem.* **255,** 2153–2159.

Lockhard, R. E. Alzner-Deweerd, B., Heckman, J. E., MacGee, J. Tabor, M. W., and RajBhandary, U. L. (1978). *Nucleic Acids Res.* **5,** 37–56.

McGeogh, D. J., and Turnbull, N. T. (1978). *Nucleic Acids Res.* **5,** 4007–4024.

Maniatis, T., and Efstradiatis, A. (1980). *In* "Methods in Enzymology" (L. Grossman and K. Moldave, eds.), Vol. 65, pp. 299–305. Academic Press, New York.

Maxam, A. M., and Gilbert, W. (1977). *Proc. Natl. Acad. Sci. U.S.A.* **74,** 560–564.

Maxam, A. M., and Gilbert, W. (1980). *In* "Methods in Enzymology" (L. Grossman and K. Moldave, eds.), Vol. 65, pp. 499–560. Academic Press, New York.

Muthukrishnan, S., and Shatkin, A. J. (1975). *Virology* **64,** 96–105.

Neu, H. C., and Heppel, L. A. (1964). *J. Biol. Chem.* **239,** 2927–2934.

Nishimura, S. (1972). *In* "Progress in Nucleic Acid Research and Molecular Biology," Vol. 12, pp. 49–85. Academic Press, New York.

Noller, H. F., and Woese, C. R. (1981). *Science* **212,** 403–411.

Peattie, D. A. (1979). *Proc. Natl. Acad, Sci. U.S.A.* **76,** 1760–1764.

Peattie, D. A. (1983a). *In* "Methods of DNA and RNA Sequencing" (S. Weisman, ed.), pp. 261–304. Praeger, New York.

Peattie, D. A. (1983b). *In* "Techniques in Nucleic Acid Biochemistry" (R. A. Flavell, ed.), Vol. B5, pp. B509/1–22. Elsevier Scientific Publishing Co., Ireland.

Peattie, D. A., and Gilbert, W. (1980). *Proc. Natl. Acad. Sci. U.S.A.* **77,** 4679–4682.

Peattie, D. A., and Herr, W. (1981). *Proc. Natl. Acad. Sci. U.S.A.* **78,** 2273–2277.

Peattie, D. A., Douthwaite, S., Garrett, R. A., and Noller, H. F. (1981). *Proc. Natl. Acad. Sci. U.S.A.* **78,** 7331–7335.

Philippsen, P., Thiebe, R., Wintermeyer, W., and Zachau, H. G. (1968). *Biochem. Biophys. Res. Commun.* **33,** 922–928.

Pochon, F., Pascal, Y., Pitha, P., and Michelson, A. M. (1970). *Biochim. Biophys. Acta* **213,** 273–281.

Ross, A., and Brimacombe, R. (1978). *Nucleic Acids Res.* **5,** 241–256.

Rubin, G. (1973). *J. Biol. Chem.* **248,** 3860–3875.

Sanger, F., and Coulson, A. R. (1978). *FEBS Lett.* **87,** 107–110.

Shatkin, A. J. (1976). *Cell (Cambridge, Mass.)* **9,** 645–653.

Silberklang, M., Proshiantz, A., Haenni, A.-L., and RajBhandary, U. L. (1977). *Nucleic Acids Res.* **4,** 4091–4108.

Silberklang, M., Gillum, A. M., and RajBhandary, U. L. (1979). *In* "Methods in Enzymology" (K. Moldave and L. Grossman, eds.), Vol. 59, pp. 58–109. Academic Press, New York.

Simoncits, A., Brownlee, G. G., Brown, R. S., Rubin, J. R., and Guilley, H. (1977). *Nature (London)* **269,** 833–836.

Simsek, M., Ziegenmeyer, J., Heckman, J., and RajBhandary, U. L. (1973). *Proc. Natl. Acad. Sci. U.S.A.* **70,** 1041–1045.

Steinschneider, A., and Fraenkel-Conrat, H. (1966). *Biochemistry* **5,** 2735–2743.

Stepanova, O. B., Metelev, V. G., Chichkova, N. V., Smirnov, V. D., Rodionova, N. P., Atabekov, J. G., Bogdanova, A. A., and Shabarov, Z. A. (1979). *FEBS Lett,* **103,** 197–199.

Tuchinskii, M. F., Gus'kova, L. I., Khazai, I., Budovskii, I., and Kochetov, N. K. (1970). *Mol. Biol. (Moscow)* **4,** 343–348.

Vanyushin, B. F., and Bur'yanov, Y. I. (1969). *Biokhimiya (Moscow)* **34,** 574–582.

Vassilenko, S., and Babkina, V. (1965). *Biokhimiya (Moscow)* **30,** 705–712.

Vincze, A., Henderson, R. E. L., McDonald, J.J., and Leonard, N. J. (1973). *J. Am. Chem. Soc.* **95,** 2677–2682.

Whitfeld, P. R. (1954). *Biochem. J.* **58,** 390–396.

Whitfeld, P. R. (1965). *Biochim. Biophys. Acta* **108,** 202–210.

Wintermeyer, W., and Zachau, H. G. (1975). *FEBS Lett.* **58,** 306–309.

Yu, C.-T, and Zamecnik, P. C. (1960). *Biochim. Biophys. Acta* **45,** 148–154.

Zimmern, D., and Kaesberg, P. (1978). *Proc. Natl. Acad. Sci. U.S.A.* **75,** 4257–4261.

APPENDIX: CHEMICAL SEQUENCING OF END-LABELED RNA MOLECULES

Outline for Four Base-Specific Reactions[a] **for Chemical Sequencing of End-Labeled RNA Molecules**

G	A > G	U	C >U[b]
300 μl G buffer	200 μl A,G buffer		
		5 μl RNA-^{32}P	5 μl RNA-^{32}P
5 μl^{32}P RNA	5 μl^{32}P RNA	Lyophilize	Lyophilize
Chill to 0°C	Chill to 0°C	10μl Hz/H$_2$O	10 μl Hz/NaCl
0.5 μl DMS	1 μl DEP	Mix, spin	Mix, spin
90°C, 0.5–1 min	90°C, 5–10 min	0°C, 5–10 min	0°C, 20–30 min
0°C	0°C	—	—
75 μl G Pptn	50 μl A,G Pptn	200 μl U Pptn	—

G	A > G	U	C >U[b]
900 μl EtOH	750 μl EtOH	600 μl EtOH	500 μl C,U Pptn
Pellet	Pellet	Pellet	Pellet
200 μl 0.3 M NaAc	200 μl 0.3 M NaAc	200 μl 0.3 M NaAc	200 μl 0.3 M NaAc
600 μl EtOH	600 μl EtOH	600 μl EtOH	600 μl EtOH
Pellet	Pellet	Pellet	Pellet
EtOH wash, dry	EtOH wash, dry	EtOH wash, dry	EtOH wash, dry
10 μl 1.0 M Tris-Cl, pH 8.2	—	—	—
10 μl 0.2 M NaBH$_4$	—	—	—
0°C, 30 min/dark	—	—	—
200 μl NaBH$_4$ Stop Mix	—	—	—
600 μl EtOH	—	—	—
Pellet	—	—	—
EtOH wash, dry	—	—	—
20 μl 1M aniline pH 4.5	20 μl 1 M aniline pH 4.5	20 μl 1 M aniline pH 4.5	20 μl M aniline pH 4.5
60°C, 20 min/dark	60°C, 20 min/dark	60°C, 20 min/dark	60°C, 20 min/dark
Lyophilize	Lyophilize	Lyophilize	Lyophilize
20 μl water	20 μl water	20 μl water	20 μl water
Lyophilize	Lyophilize	Lyophilize	Lyophilize
20 μl water	20 μl water	20 μl water	20 μl water
Lyophilize	Lyophilize	Lyophilize	Lyophilize
2–3 μl loading buffer	2–3 μl loading buffer	2–3 μl loading buffer	2–3 μl loading buffer
90°C, 30 sec	90°C, 30 sec	90°C, 30 sec	90°C, 30 sec
Load on gel	Load on gel	Load on gel	Load on gel

[a] Reagents and procedures for the reactions are given in the notes below.

[b] An alternative C + U reaction is described in Section III,A.

Notes to the Appendix

1. G Reaction

G Buffer:

> 50 mM sodium cacodylate-HCl (pH 5.5)
> 1 mM EDTA

Do not autoclave this buffer. The pH will drop drastically, and the poisonous and volatile compound cacodyl forms upon heating.

G Pptn solution:

> 1.0 M Tris-acetate (pH 7.5)
> 1.0 M 2-mercaptoethanol
> 1.5 M sodium acetate

0.1 mM EDTA
0.2 mg/ml carrier tRNA

NaBH$_4$ stop:

0.6 M sodium acetate (pH 4.5)
0.6 M acetic acid (pH 4.5)
0.025 mg/ml carrier tRNA

Note well that the NaBH$_4$ solution *must* be fresh.

2. A > G Reaction

A,G Buffer:

50 mM sodium acetate to pH 4.5 with acetic acid
1 mM EDTA

A,G Pptn solution:

1.5 M sodium acetate
0.2 mg/ml carrier tRNA

3. U Reaction

Hz/H$_2$O:

Anhydrous hydrazine/H$_2$O (1 : 1) fresh and cooled to 0°C.

U Pptn solution:

0.3 M sodium acetate
0.1 mM EDTA
0.05 mg/ml carrier tRNA

4. C > U Reaction

Hz/NaCl:

Oven-dried NaCl to 3.0 M in anhydrous hydrazine, fresh and cooled to 0°C.

C,U Pptn solution:

Ethanol/H$_2$O (4 : 1)

8

Oligonucleotide-Directed Site-Specific Mutagenesis Using Double-Stranded Plasmid DNA

SUMIKO INOUYE[1]
MASAYORI INOUYE[1]
Department of Biochemistry
State University of New York at Stony Break
Stony Brook, New York 11794

I. INTRODUCTION

Introduction of a specific mutation at a specific site of a known gene is a most powerful way to study various aspects of the gene including the mechanism of gene regulation, the expression of the gene product, and structural and functional analysis of the gene product. Restriction enzyme cleavage sites found in the gene can be used for site-specific mutagenesis. However, oligonucleotide-directed site-specific mutagenesis is far more versatile and is the most effective method to create desired mutations, such as base substitutions, deletions, and insertions, at a desired site in a specific gene.

This method was first developed to alter the DNA sequence of the single-stranded DNA bacteria phage ϕX174 (Razin *et al.*, 1978; Hutchinson *et al.*, 1978; Gillam and Smith, 1979a,b). The method involves the use of a short, synthetic oligodeoxyribonucleotide carrying the appropriate

---

[1] Present address: Department of Biochemistry, Robert Wood Johnson Medical School, University of Medicine and Dentistry of New Jersey, Piscataway, New Jersey 08854-5635.

Synthesis and Applications
of DNA and RNA

mutation, which acts as a primer of *in vitro* DNA synthesis on a complementary single-stranded circular DNA template. Following transformation of competent cells with the resulting heteroduplex containing the incorporated oligodeoxyribonucleotide, *in vivo* semiconservative replication resolves the heteroduplex into homoduplexes derived from mutant and parental strands. The versatility of this method has been greatly enhanced by use of double-stranded recombinant plasmids as the source of the single-stranded templates and by use of improved colony-hybridization techniques. Using the ^{32}P-labeled synthetic oligodeoxyribonucleotide mutagen as the probe allows the isolation of site-directed mutants without phenotypic screening (Wallace *et al.*, 1980).

Mutagenesis through synthetic oligodeoxyribonucleotides has been used to introduce deletions, insertions, transitions, and transversions at specific sites within a given DNA sequence (Gillam and Smith, 1979a,b; Gillam *et al.*, 1979, 1980; Kudo *et al.*, 1981; Zoller and Smith, 1982; Montell *et al.*, 1982; Simons *et al.*, 1982). In addition, the use of this method in the creation of mutant proteins with specific amino acid changes has allowed precise investigations of structure–function relationships (Wallace *et al.*, 1981b; Inouye *et al.*, 1982; Charles *et al.*, 1982; Dalbadie-McFarland *et al.*, 1982; Winter *et al.*, 1982). There is no doubt that this technology will play a major role in protein engineering, which includes not only studies on various aspects of proteins but also the creation of new enzymes with different structures and different or new specificities.

We have used oligonucleotide-directed mutagenesis extensively to create specific mutations within the signal peptide region of the major outer-membrane lipoprotein of *Escherichia coli* to elucidate the function of this N-terminal extension consisting of 20 amino acid residues in the secretion of the lipoprotein across the cytoplasmic membrane (Inoye *et al.*, 1982, 1983a,b, 1984; Vlasuk *et al.*, 1983; 1984). These examples are summarized in Fig. 1, where various amino acid substitutions, insertions, and deletions obtained by the oligonucleotide-directed site-specific mutagenesis are listed. Similarly, many researchers have started to apply this technology to a variety of enzymes and protein factors in order to study their structures and functions: human β-globin (Wallace *et al.*, 1981b), tyrosyl-tRNA synthesis (Winter *et al.*, 1982; Wilkinson *et al.*, 1984), β-lactamase (Charles *et al.*, 1982; Dalbadie-McFarland *et al.*, 1982), dihydrofolate reductase (Villafranca *et al.*, 1983), polyoma virus middle-T antigen (Oostra *et al.*, 1983), SV40 large-T antigen (Lewis *et al.*, 1983), T4 phage lysozyme (Perry and Wetzel, 1984), human interleukin-2 (Wang *et al.*, 1984), ribulose-1,5-bisphosphate carboxylase (Gutteridge *et al.*, 1984), the human c-Ha-*ras*1 gene product (Seeburg *et al.*, 1984), α_1-antitrypsin (Courtney *et al.*, 1985), and yeast cytochrome *c* (Pielak, *et al.*, 1985). Applications of the site-specific mutagenesis to regulatory regions of

```
                            1           5              10             15            20▼
Wild Type                   Met-Lys-Ala-Thr-Lys-Leu-Val-Leu-Gly-Ala-Val-Ile-Leu-Gly-Ser-Thr-Leu-Leu-Ala-Gly-Cys-

Mutant I1   (D3)            Met-Lys-(Asp)-Thr-Lys-Leu-Val-Leu-Gly-Ala-Val-Ile-Leu-Gly-Ser-Thr-Leu-Leu-Ala-Gly-

       I2   (Δ2)            Met-(  )-Ala-Thr-Lys-Leu-Val-Leu-Gly-Ala-Val-Ile-Leu-Gly-Ser-Thr-Leu-Leu-Ala-Gly-

       I3   (Δ2,D3)         Met-(  )-(Asp)-Thr-Lys-Leu-Val-Leu-Gly-Ala-Val-Ile-Leu-Gly-Ser-Thr-Leu-Leu-Ala-Gly-

       I4   (E2,D3)         Met-(Glu)-(Asp)-Thr-Lys-Leu-Val-Leu-Gly-Ala-Val-Ile-Leu-Gly-Ser-Thr-Leu-Leu-Ala-Gly-

       I5   (N5)            Met-Lys-Ala-Thr-(Asn)-Leu-Val-Leu-Gly-Ala-Val-Ile-Leu-Gly-Ser-Thr-Leu-Leu-Ala-Gly-

       I6   (Δ2,N5)         Met-(  )-Ala-Thr-(Asn)-Leu-Val-Leu-Gly-Ala-Val-Ile-Leu-Gly-Ser-Thr-Leu-Leu-Ala-Gly-

       I7   (E2,D3,N5)      Met-(Glu)-(Asp)-Thr-(Asn)-Leu-Val-Leu-Gly-Ala-Val-Ile-Leu-Gly-Ser-Thr-Leu-Leu-Ala-Gly-


       A1   (V9)            Met-Lys-Ala-Thr-Lys-Leu-Val-Leu-(Val)-Ala-Val-Ile-Leu-Gly-Ser-Thr-Leu-Leu-Ala-Gly-

       A2   (Δ9)            Met-Lys-Ala-Thr-Lys-Leu-Val-Leu-(  )-Ala-Val-Ile-Leu-Gly-Ser-Thr-Leu-Leu-Ala-Gly-

       B1   (V14)           Met-Lys-Ala-Thr-Lys-Leu-Val-Leu-Gly-Ala-Val-Ile-Leu-(Val)-Ser-Thr-Leu-Leu-Ala-Gly-

       B2   (Δ14)           Met-Lys-Ala-Thr-Lys-Leu-Val-Leu-Gly-Ala-Val-Ile-Leu-(  )-Ser-Thr-Leu-Leu-Ala-Gly-

     A1B1   (V9,V14)        Met-Lys-Ala-Thr-Lys-Leu-Val-Leu-(Val)-Ala-Val-Ile-Leu-(Val)-Ser-Thr-Leu-Leu-Ala-Gly-

     A1B2   (V9,Δ14)        Met-Lys-Ala-Thr-Lys-Leu-Val-Leu-(Val)-Ala-Val-Ile-Leu-(  )-Ser-Thr-Leu-Leu-Ala-Gly-

     A2B1   (Δ9,V14)        Met-Lys-Ala-Thr-Lys-Leu-Val-Leu-(  )-Ala-Val-Ile-Leu-(Val)-Ser-Thr-Leu-Leu-Ala-Gly-

     A2B2   (Δ9,Δ14)        Met-Lys-Ala-Thr-Lys-Leu-Val-Leu-(  )-Ala-Val-Ile-Leu-(  )-Ser-Thr-Leu-Leu-Ala-Gly-


  A2B2α1    (Δ7,Δ9,Δ14)     Met-Lys-Ala-Thr-Lys-Leu-(  )-Leu-(  )-Ala-Val-Ile-Leu-(  )-Ser-Thr-Leu-Leu-Ala-Gly-

  A2B2β1    (Δ9,Δ13,Δ14)    Met-Lys-Ala-Thr-Lys-Leu-Val-Leu-(  )-Ala-Val-Ile-(  )-(  )-Ser-Thr-Leu-Leu-Ala-Gly-

A2B2α1β1    (Δ7,Δ9,Δ13,Δ14) Met-Lys-Ala-Thr-Lys-Leu-(  )-Leu-(  )-Ala-Val-Ile-(  )-(  )-Ser-Thr-Leu-Leu-Ala-Gly-


       A3   (D9)            Met-Lys-Ala-Thr-Lys-Leu-Val-Leu-(Asp)-Ala-Val-Ile-Leu-Gly-Ser-Thr-Leu-Leu-Ala-Gly-

       B3   (D14)           Met-Lys-Ala-Thr-Lys-Leu-Val-Leu-Gly-Ala-Val-Ile-Leu-(Asp)-Ser-Thr-Leu-Leu-Ala-Gly-

       A4   (R9)            Met-Lys-Ala-Thr-Lys-Leu-Val-Leu-(Arg)-Ala-Val-Ile-Leu-Gly-Ser-Thr-Leu-Leu-Ala-Gly-


       H1   (A15)           Met-Lys-Ala-Thr-Lys-Leu-Val-Leu-Gly-Ala-Val-Ile-Leu-Gly-(Ala)-Thr-Leu-Leu-Ala-Gly-

       H2   (A16)           Met-Lys-Ala-Thr-Lys-Leu-Val-Leu-Gly-Ala-Val-Ile-Leu-Gly-Ser-(Ala)-Leu-Leu-Ala-Gly-

       H3   (A15,A16)       Met-Lys-Ala-Thr-Lys-Leu-Val-Leu-Gly-Ala-Val-Ile-Leu-Gly-(Ala)-(Ala)-Leu-Leu-Ala-Gly-


       C2   (G21)           Met-Lys-Ala-Thr-Lys-Leu-Val-Leu-Gly-Ala-Val-Ile-Leu-Gly-Ser-Thr-Leu-Leu-Ala-Gly-(Gly)-

       C1   (A20)           Met-Lys-Ala-Thr-Lys-Leu-Val-Leu-Gly-Ala-Val-Ile-Leu-Gly-Ser-Thr-Leu-Leu-Ala-(Ala)-

       C3   (Δ20)           Met-Lys-Ala-Thr-Lys-Leu-Val-Leu-Gly-Ala-Val-Ile-Leu-Gly-Ser-Thr-Leu-Leu-Ala-(  )-

       C5   (Δ18)           Met-Lys-Ala-Thr-Lys-Leu-Val-Leu-Gly-Ala-Val-Ile-Leu-Gly-Ser-Thr-Leu-(  )-Ala-Gly-

       C6   (S20)           Met-Lys-Ala-Thr-Lys-Leu-Val-Leu-Gly-Ala-Val-Ile-Leu-Gly-Ser-Thr-Leu-Leu-Ala-(Ser)-

       C7   (V20)           Met-Lys-Ala-Thr-Lys-Leu-Val-Leu-Gly-Ala-Val-Ile-Leu-Gly-Ser-Thr-Leu-Leu-Ala-(Val)-
```

```
Mature Region
                            20▼
Wild Type                   Gly-Cys-Ser-Ser-Asn-Ala-Lys

       M4   (I23,I24)        Gly-Cys-Ser-(Ile)-(Ile)-Ala-Lys

     M4C1   (A20,I23,I24)    (Ala)-Cys-Ser-(Ile)-(Ile)-Ala-Lys
```

Fig. 1. Mutants of the major outer membrane lipoprotein of *Escherichia coli*. All mutants were isolated by oligonucleotide-directed site-specific mutagenesis using double-stranded plasmid DNAs as templates. The amino acid residues mutated are circled. Deletion mutations are shown by empty circles. Arrows indicate the position of the signal peptide cleavage site.

of genes such as promoters, enhancers, 5'-end noncoding regions, ribosome-binding sites, transcription termination sites, splicing sites, etc., are also very important and have been reported (see review by Itakura *et al.*, 1984).

In this article, we will describe the method developed in our laboratory using double-stranded plasmid DNA as a template rather than single-stranded M13 DNA. Those who are interested in the method for oligonucleotide-directed site-specific mutagenesis with M13 vectors should refer to an excellent step-by-step protocol written by Zoller and Smith (1983).

II. GENERAL CONSIDERATIONS

A. How to Choose the Template DNA for Mutagenesis: Double-Stranded Plasmid DNA or Single-Stranded M13 DNA

As mentioned in the Introduction, there are two basic methods for site-directed mutagenesis using synthetic oligonucleotides as mutagens; one utilizes double-stranded plasmid DNA as a template (the plasmid method) and the other utilizes single-stranded M13 DNA as a template (the M13 method). Therefore, it is important to know the shortcomings and advantages of each method in order to choose which method should be used for the mutagenesis of a particular gene to be mutagenized. In the following sections, the unique features of the plasmid method are discussed and compared to the M13 method.

1. Double-Stranded DNA as Template

In most cases, a gene of interest is initially cloned in a plasmid vector. On the other hand, with the M13 method, an appropriate DNA fragment must be first subcloned into an M13 vector and, after mutagenesis, it usually must be recloned back into a plasmid vector to examine the biological effect of the mutagenesis. These subcloning steps are eliminated in the plasmid method.

2. Size Limitation of the Template DNA

The size of DNA fragments that can be cloned in an M13 vector is more limited than in a plasmid vector. In general, DNA fragments larger than 2 kilobases (kb) in the M13 vectors are unstable and can be lost during phage propagation (Cordell *et al.*, 1979; Barnes, 1980).

3. Availability of Restriction Sites

For both methods, a smaller-size target DNA is desirable to avoid the occurrence of nonspecific mutagenesis at sites besides the one to be mutagenized. In the plasmid method, two unique restriction enzyme cleavage sites are required to create a single-stranded region (see Fig. 2). In the M13 method, the sites flanking the target DNA need not necessarily be unique. However, in the latter case, the use of restriction enzymes which create blunt ends upon cleavage may prevent recovery of the cloned DNA fragment from the M13 clone as the identical fragment.

4. Confirmation of the Mutation

After mutagenesis, it is most desirable to confirm the mutation by DNA sequencing. In the case of the M13 method, the dideoxy sequencing method (Messing *et al.*, 1981) has been well established using a universal primer. The universal primer can be conveniently used if the mutation is close to the cloning site. Otherwise, another synthetic oligonucleotide or a restriction fragment has to be used as a primer (Zoller and Smith, 1983). In the case of the plasmid method, an appropriate restriction site close to the mutation site should be available for end labeling with ^{32}P. DNA sequencing is then carried out according to the method of Maxam and Gilbert (1980). Dideoxy sequencing is also possible directly on double-stranded plasmid DNAs by using a specific DNA fragment as a primer (Wallace *et al.*, 1981a).

5. Mutant Yields

The mutant yields by the M13 method are generally higher than those by the plasmid method for the following two reasons: (a) In the plasmid method, only one of the two heteroduplex template DNAs (DNA-a and -b in Fig. 2) can create a mutant. Therefore, the mutant yield by this method cannot be higher than 50%, in contrast to 100% possible with the M13 method. (b) In the plasmid method, a fraction of the ampicillin-resistant transformants after the mutagenesis may be derived from homoduplex DNA fragments (DNA I and II in Fig. 2). These background colonies contribute to the lower mutant yield. In practice, however, the mutant yields by the plasmid method described in this chapter are usually 3–15% and can be as high as 25% (see Table I), which are comparable to those obtained by the M13 method [Table I in the article by Zoller and Smith (1983)].

6. Other Considerations

The existence of two complementary template DNAs in the plasmid method (DNA-a and -b, in Fig. 2) are advantageous in the following cases: (a) Two independent synthetic oligonucleotides can be used to independently create different mutations on the individual template plasmid DNAs at the same time, if one oligonucleotide is designed for one strand and the other for the complementary strand. (b) Some mutations may be

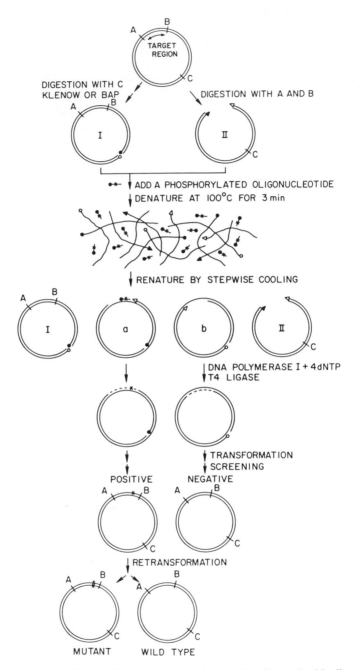

Fig. 2. Schematic diagram depicting the steps involved in oligonucleotide-directed site-specific mutagenesis of a cloned circular double-stranded plasmid DNA. Experimental details are described in the text.

TABLE I

Examples of Oligonucleotide-Directed Mutagenesis Using the Plasmid DNAs as Templates

Case	Sequence[a]	Mutation	Length of oligonucleotide	Size of plasmid (kb)	Length of single-stranded gap (kb)	Mutant yield (%)
1	5'TTTGTGTTATACTTG3'	A → T	15	7.4	1.0	3.1
2	TATGTTGTGAAATATT	A → T	15	7.4	1.0	5.0
3	TTATGTTGTCAATATTTT	AG → TC	18	7.4	1.0	11.7
4	TTGTGTTATAATTGTAACG	C → A	19	7.4	1.0	25.5
5	TTGTGTTATAATTGTAACG	A → T, C → A	19	7.4	1.0	16.3
6	TTGAGTTAGATCTCCAT	G-insertion	17	7.4	1.0	4.0
7	ATCCTGGATTCTACT	G → A	15	5.5	0.56	3.5
8	CTGGCAAGTTGCTCC	G → A	15	5.5	0.56	3.7
9	CTGGCAGTTTGCTCC	G → T	15	5.5	0.56	3.5
10	ACGCTAATATCGATC	A → T	15	5.5	0.56	13.0
11	TGCTCCATCATCGCTAAA	G → T, A → T	18	5.5	0.56	3.2
12	CAGCCATTTAAATAACGGG	GCC → TAA	19	5.7	0.37	3.1
13	TCGAAGAGCTCAACGCCA	TG → CT	18	5.7	0.37	2.1
14	GCGGTAATCTCTACTCTG	CTG-deletion	18	5.5	0.56	25.0
15	ACTAAACTGCTGGCGGTA	GTA-deletion	18	5.5	0.56	1.5
16	GGCACTGCCGGGGTTTCGC	G → C, T → G	18	5.5	0.56	7.0
17	TTTCGCTGCAGTAGCGCA	A → G, C → A	18	7.4	0.081	4.0
18	AAAAAATGAGCGCTATCGCG	AAAAGACA → GC	20	7.4	0.081	2.0
19	GTAGCGCAGGCCCACCCAGAAACG	12-base deletion	24	12.0	0.57	2.5
20	TTAATAATGAAGCTTCTGGCGGTA	9-base deletion A → G, G → T	24	5.5	0.56	2.5

[a] Mutated bases are indicated by dots. Arrows indicate the position where deletion mutations occurred.

187

very difficult to obtain because of the secondary structure around the target site and/or because of the distance between the target site and the double-stranded region. In such a case, the use of the oligonucleotide complementary to the other DNA strand often dramatically improves the frequency of mutagenesis.

In the M13 method, the closed-circular DNA is generally purified by alkaline sucrose density gradient centrifugation (Zoller and Smith, 1983). This laborious enrichment procedure is not used in the plasmid method.

Another important advantage of the plasmid method is that the size of the single-stranded region can be confined to a small region around the target site. With the M13 method, the entire vector (which is larger than 7 kb) is single stranded. This greatly increases the possibility that the oligonucleotide primer will induce mutations at nontargeted single-stranded sites. Mutations outside the target region would give rise to false positives whose identity can only be revealed by nucleotide sequence determination.

B. Outline of the Plasmid Method

The general scheme for the plasmid method is illustrated in Fig. 2. In this method, only the desired region to be mutagenized is converted to the single-stranded form. As a result, substantial improvement in the mutant yield has been achieved in comparison to the original plasmid method. In the original method, supercoiled plasmid DNA carrying a target gene was first converted to open circular DNA by partial restriction endonuclease digestion in the presence of ethidium bromide. The open-circular DNA was then treated with exonuclease III, followed by annealing with a synthetic oligonucleotide (Wallace *et al.*, 1980, 1981b; Vlasuk and Inouye, 1983). This created single-stranded regions not only at the target site but also at various other sites, resulting in a very poor mutant yield (usually less than 1%). However in the new method depicted in Fig. 2, the single-stranded region is created only at the target site, which dramatically simplifies the procedure and increases the mutant yield (Morinaga *et al.*, 1984). A similar method was first reported for the isolation of mutants of polyama virus middle-T antigen by Oostra *et al.* (1983). The following are step-by-step procedures for mutagenesis using plasmid DNA.

1. Preparation of Plasmid DNA Fragments I and II (Step I)

A plasmid DNA carrying a gene of interest has three restriction sites, A, B, and C (A and B can be for the same enzyme). It is desirable for the C site to be in the drug-resistance gene of the plasmid, and the distance between the A and B sites is preferably less than 2 kb.

One-half of the plasmid DNA is digested at C, followed by treatment with the Klenow fragment of DNA polymerase I, or nuclease S_1, or bacterial alkaline phosphatase (BAP). The Klenow or nuclease S_1 treatment removes single-stranded sequences resulting from restriction endonuclease digestion at the site C. The resulting fragment I cannot confer drug resistance even if it forms a closed circular plasmid upon ligation, provided that site C is in a drug-resistant gene. For example, the unique *Pst*I site in the ampicillin-resistant gene of pBR322, can be used as the C site. If C site digestion results in blunt cleavage or is not in a drug-resistance gene, BAP treatment is essential to prevent self-ligation of fragment I. The C site should be at least several hundred base pairs from both the A and B sites. The C site need not be unique if the gap produced by C digestion is not too large and does not overlap with the gap produced by A plus B digestion.

The other half of plasmid DNA is digested at A and B to remove the target region. A small target region is desirable for higher mutant yield and for reducing the probability of the heteroduplex formation with the mutagenic oligonucleotide at an alternative site rather than the target site. A reasonable mutant yield, however, can be obtained even with a single-stranded region as long as 2 kb. The resulting fragment II is purified by 5% polyacrylamide gel electrophoresis. If A and B are the same enzyme, BAP treatment should be performed again to prevent self-ligation of fragment II.

If the target region is too large because of the lack of appropriate restriction sites, a smaller fragment may be isolated from the fragment AB, which is then added back in the denaturation mixture. Using this strategy, the target site can still be kept in a single-stranded region of an appropriate size as discussed in Section IV.

2. Denaturation and Renaturation (Step II)

Equimolar amounts of fragments I and II are mixed with a 100- to 500-fold molar excess of synthetic oligonucleotide. The mixture is incubated at 100°C for 3 min to denature completely the DNA fragments in the mixture. After the denaturation, the mixture is gradually cooled in a stepwise manner to allow the denatured DNA fragments to reanneal. During this reannealing procedure, two new species of DNAs, DNA-a and -b are formed in addition to the original fragments I and II. As shown in Fig. 2, the mutagenic oligonucleotide hybridizes with only one of the two species of circular DNAs to form the heteroduplex, since the oligonucleotide is complementary to only one of the two strands of the target gene.

The formation of the new circular DNAs should be confirmed by agarose gel electrophoresis before proceeding to the next step. As shown in Fig. 3, after reannealing a new band clearly appears at the position indicated by arrow a + b. This slower-migrating band consists of the new circular DNAs, DNA-a and DNA-b, which are structurally almost identi-

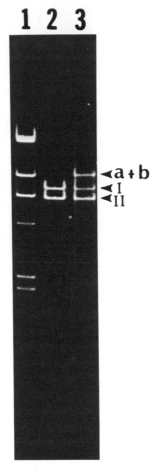

Fig. 3. Analysis of DNA after denaturation followed by renaturation. Fragments I and II were prepared from the pIN-III-A3 vector as described in Section III,C and D. Lane 1, *Hin*dIII digest of bacteriophage λ DNA as a molecular weight standard. Their molecular weights are 23.1, 9.4, 6.6, 4.4, 2.3, and 2.0 kilobases from the top to the bottom band, respectively; lane 2, the mixture of fragments I and II; lane 3, the same mixture was denatured at 100°C for 3 min followed by gradual cooling as described in Section III,E. Gel electrophoresis was carried out with use of 0.7% agarose in 40 m*M* Tris–20 m*M* sodium acetate (pH 8.0) containing 2 m*M* EDTA. Bands a + b, I, and II correspond to DNA-a and -b, and fragments I and II shown in Fig. 2, respectively.

cal. The position of the new band is dependent upon the size of the single-stranded region of the new circular DNAs. Sometimes, the new circular DNAs migrate at the same position as fragment I. In this case, the ratio of the intensity of band I to that of band II increases after reannealing. The migration of the new band becomes faster as the size of the single-stranded region increases. The open circular DNA also migrates faster when gel electrophoresis is carried out at a lower voltage for a longer time.

3. Primer Extension and Transformation (Step III)

After confirming the formation of the heteroduplex DNAs, the DNAs are incubated with the Klenow fragment of DNA polymerase I, T4 ligase, and four deoxyribonucleotide triphosphates. This treatment converts the open circular DNAs to closed circular DNAs. (When fragment I is prepared by BAP treatment in Step I, they still remain open circular.) In the case of DNA-a, the oligonculeotide serves as a primer for the polymerase. After the reaction, the mixture is used for transformation.

When A and B are the same restriction sites, the addition of ligase can be eliminated to avoid self-ligation of fragment II. The elimination of ligase reduces the number of transformants by a factor of 2–3 and reduces the mutant yield if A and B are different, but increases the mutant yield if A and B are the same. In this regard, it is interesting to note that mutation can be obtained without Klenow and ligase treatment, although the mutant yield is very low (approximately 0.5%; Y. Mukohara and S. Inouye, unpublished results). This indicates that the open circular DNA, DNA-a and DNA-b, can be incorporated into the cells, and then repaired to form the closed-circular DNA.

4. Screening and Confirmation of the Mutations (Step IV)

Transformants carrying the mutated plasmid are screened by colony hybridization using the same synthetic oligonucleotide which was used as the mutagen. The oligonucleotide labeled with ^{32}P is used as the probe for the screening. It forms an unstable heteroduplex with the wild-type plasmid DNA, but forms a perfectly complementary stable homoduplex with the mutant plasmid DNA. Therefore, the hybridization temperature is chosen for each oligonucleotide so that stable duplexes are formed only with the mutant but not with the wild-type plasmid DNA.

The colony hybridization can be performed by three different methods as shown in Fig. 4. The clearest and surest result is usually obtained by the pick-and-screen method (Fig. 4A), where each transformant is picked and transferred onto a filter paper (Whatman 3MM). However, when it is necessary to screen many transformants, they can be transferred onto

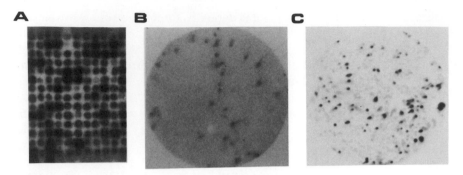

Fig. 4. Colony-hybridization analysis using ^{32}P-labeled oligonucleotide as the probe. (A) Autoradiogram of individual transformants which were picked, and transferred onto a filter paper (Whatman 3 MM) as described in Section III,H (Pick-and-Screen Method). (B) Autoradiogram of a paper disk (Whatman No. 1) containing initial transformants (Filter Assay Method I). (C) Autoradiogram of nitrocellulose filter disk containing initial transformants (Filter Assay Method II). Hybridization was carried out as described in Section III, I.

either a round Whatman No. 1 filter paper (filter-assay method I; Fig. 4B) or a nitrocellulose filter (filter-assay method II; Fig. 4C) directly from a plate. Method B is less expensive but positive spots obtained by method B are not as sharp as those obtained by method C.

Since each positive transformant may contain the mutant plasmid as well as the wild-type plasmid, it is important to isolate the mutant plasmid by retransformation. From one or two retransformed clones, plasmid DNA is purified and sequenced to confirm the mutation caused by the mutagenic oligonucleotide.

If a restriction site is lost or created as a result of the mutation, restriction site analysis of the plasmid DNAs from the transformants can also be used for the primary screening for the mutant.

C. Design of Mutagenic Oligonucleotides

For a base substitution mutation, an oligonucleotide of 15 bases in length (15-mer) is usually used, in which the mismatched base is in the center. In general, seven to nine bases from the mutation site appear to be sufficient. Various examples are shown in Table I along with the sizes of plasmid DNAs, the sizes of the single-stranded regions, and the mutant yields. Table I lists not only various one-base mismatches, but also two-base mismatches where the two bases are separated by 0 (cases 3 and 13), 1 (cases 17 and 18), 2 (case 11), and 3 bases (case 5). Table I also includes a 3-base mismatch (case 12), two 3-base deletions (cases 14 and 15), a deletion of 12 bases (cases 19), a substitution of an 8-base sequence for 2 bases (case 18), and a 9-base deletion flanked by 2-base mismatches (case

20). In designing the oligonucleotides, it is sometimes possible to create a desirable mutation together with a new restriction site or to eliminate a restriction site at the target site. Such a strategy is useful in screening for the mutant as well as for manipulation of the gene. To avoid the hybridization of the oligonucleotide with any other regions besides the target site, the DNA sequence of the single-stranded region should be examined for possible complementarity with the oligonucleotide. If such a region is found, a longer oligonucleotide should be synthesized which has more mismatches between the oligonucleotide and the partially complementary region.

III. EXPERIMENTAL PROCEDURES

A. Preparation of Oligonucleotide Primer

Oligonucleotides are purified by polyacrylamide gel electrophoresis (Vlasuk and Inouye, 1983) and dissolved in sterile 0.1 × TE to a concentration of 20 pmole/μl.

1. Solution

TE: 10mM Tris-HCl buffer (pH 7.5), 1 mM EDTA
TEAB: Triethylamine carbonate buffer (pH 8.0). A stock solution (2 M) is prepared by dissolving CO_2 gas into a triethylamine (distilled) solution to pH 8.0.

2. Preparation of 5′-Phosphorylated Primer

1. Mix the solutions shown in the following tabulation in the order listed:

Component	Volume
H_2O	Appropriate amount to make a final volume of 50 μl
0.5 M Tris-HCl (pH 7.5)	6.6 μl
0.1 M MgCl$_2$	5 μl
0.1 M 2-Mercaptoethanol	5 μl
Oligonucleotide (20 pmole/μl)	25 μl
1 mM ATP	1.5 μl
T4 polynucleotide kinase	5 U

2. Incubate the mixture at 37°C for 60 min.

3. Stop the reaction by incubating the mixture at 65°C for 10 min. Store the mixture at −20°C.

3. Remarks

1. The phosphorylation of the oligonucleotide at the 5' end is essential for ligation. The efficiency of the phosphorylation can be routinely checked as follows: 1 μCi of [γ-^{32}P] ATP is added to the reaction mixture described above and the ^{32}P-labeled oligonucleotide is purified as described for the hybridization probe (see the next section). Under the conditions used above, approximately 30% of the total radioactivity should be incorporated into the oligonucleotide, (the theoretical maximum efficiency is 33.3%). A low efficiency indicates poor quality of the oligonucleotide.

2. When the oligonucleotide is labeled with ^{32}P as above, and subsequently purified by column chromatography [see Section III, B, 4], the fraction from the column must be lyophilized to reduce the volume. The lyophilized material is then solubilized in 50 μl of 0.1 × TE.

B. Preparation of Oligonucleotide Hybridization Probe

1. Mix the solutions shown in the following tabulation in the order listed:

Component	Volume
H$_2$O	Appropriate amount to make a final volume of 50 μl
0.5 M Tris-HCl (pH 7.5)	6.6 μl
0.1 M MgCl$_2$	5 μl
0.1 M 2-Mercaptoethanol	5 μl
Oligonucleotide (20 pmole/μl)	1.5 μl
[γ-^{32}P]ATP (3000 Ci/mmole) in H$_2$O	100 μCi
T4 polynucleotide kinase	10 U

2. Incubate the mixture at 37°C for 60 min.
3. Chill the mixture on ice.
4. Apply the mixture on a Sephadex G-50 superfine column equilibrated with 50 mM TEAB or 50 mM ammonium bicarbonate (pH 7.8). The column is prepared using a disposable plastic column (0.6 × 20 cm) with siliconized glass wool at the bottom.
5. Elute the phosphorylated oligonucleotide with the same buffer used

for equilibration of the column. The elution of the oligonucleotide should be monitored with a hand-held Geiger counter.

6. Collect the first radioactive peak into a sterile 15-ml polypropylene centrifuge tube (approximately 1.5 ml).

7. To the eluate, add EDTA to a final concentration of 1 mM and store the mixture at $-20°C$.

Remarks

1. The efficiency of ^{32}P-incorporation into the oligonucleotide should be 80–90%.

2. If necessary, the purity of the phosphorylated oligonucleotide can be examined by PEI cellulose (Brinkmann) thin-layer chromatography in a 0.75 M potassium phosphate buffer (pH 3.7) system. The phosphorylated oligonucleotide hardly migrates from the origin, whereas [γ-^{32}P]ATP and ^{32}P$_i$ migrate much faster (the R_f of ATP is ~0.5 and the R_f of P$_i$ is ~0.8).

3. At step 1, if 50 μCi [γ-^{32}P]ATP is used instead of 100 μCi, most of the ^{32}P can be incorporated into the oligonucleotide and easily examined by TLC as described above. If very little radioactivity is found in the ATP and P$_i$ fractions, steps 4–7 can be omitted. The specific activity of the probe using this method is lower than the probe labeled by step 1–7.

C. Preparation of Fragment I (Step I)

As discussed earlier, a unique restriction site should be chosen to linearize the plasmid DNA (see Fig. 2). How to choose such a site is described in Section II,B,1. In this section, we describe an example using an expression cloning vector, pIN-III-A3 (Masui *et al.*, 1984). The mutagenesis is carried out in the *lpp* promoter region in the vector. Figure 5 shows the restriction map of the vector. For the C site to create fragment I, the unique *Hpa*I site in the *lacI* gene is used. For the A and B sites to create fragment II, the *Pst*I and *Hind*III sites, respectively, are used. If the target region is between *Xba*I and *Eco*RI (if a target DNA is cloned between the two sites), the *Pst*I site in the ampicillin-resistant gene can be used as the C site (Morinaga *et al.*, 1984). Readers should refer to the manual written by Maniatis *et al.* (1982) for the various procedures of DNA manipulation used in this article.

Example

Five micrograms of pIN-III-A3 (7.4 kb) DNA are digested with *Hpa*I. After extraction with phenol (twice) and with ether (three times), the DNA is precipitated with ethanol. The DNA is solubilized in 200 μl of 10

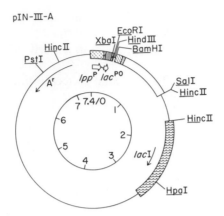

Fig. 5. Restriction map of the high-expression cloning vector, pIN-III-A3 (Masui *et al.,*
1984).

m*M* Tris-HCl (pH 7.5) containing 0.1 m*M* EDTA. Bacterial alkaline phosphatase (BAP) is added to the solution (0.3 unit) and the mixture is incubated at 65°C for 60 min. The DNA is then extracted four times with
phenol, three times with ether, and precipitated with ethanol. The precipitate is dissolved in 40 µl of 0.1 × TE. The DNA concentration of the
solution should be estimated by applying an aliquot (1 µl) an agarose
(0.7%) gel.

D. Preparation of Fragment II (Step I)

Example

Five micrograms of pIN-III-A3 are digested with *Pst*I and *Hind*III.
Fragment II (6.4 kb) is separated from the 1.0-kb fragment by 5% polyacrylamide gel electrophoresis and the subsequent electroelution. The
DNA solution is extracted with phenol, and then ether, and the DNA is
then precipitated with ethanol. The precipitate is solubilized in 40 µl of
0.1 × TE. The DNA concentration of the solution is estimated as described for fragment I.

E. Heteroduplex Formation (Step II)

1. Example

1. Mix the solutions shown in the following tabulation in a 1.5-ml
Eppendorf centrifuge tube in the order listed:

Component	Volume
H_2O	To make a final volume of 34.8 μl
Fragment I	0.3 μg
Fragment II	0.26 μg
5 × Polymerase, ligase buffer	12 μl
5'-Phosphorylated primer from III-A	3.75 μl

5 × Polymerase-Ligase Buffer

    500 *M* NaCl
    32.5 m*M* Tris-HCl (pH 7.5)
    40   m*M* MgCl$_2$
    5   m*M* 2-Mercaptoethanol

2. Remove 11.6 μl, and directly apply to an agarose gel without boiling alongside the sample from step 8 below.

3. Incubate the remaining mixture for 3 min in a boiling water bath.

4. Quickly transfer the tube to a 30°C incubator, and incubate at 30°C for 30 min.

5. Transfer the tube to a refrigerator (4°C) and incubate for 30 min.

6. Keep the mixture on ice for 10 min.

7. Spin the tube for 2 sec in an Eppendorf centrifuge at 4°C.

8. Remove 11.6 μl from the mixture and apply it to 0.7% agarose gel with the sample from 2 above.

9. Subject the samples to electrophoresis at 6 volts/cm for 2–3 hr with E buffer.

10. Stain the gel with ethidium bromide.

E Buffer (pH 8.0)

    40 m*M* Tris-acetate
    20 m*M* Sodium acetate
    2 m*M* EDTA

2. Remarks

1. As shown in Fig. 3, as a result of denaturation and renaturation, a new band at (a + b) should appear in addition to fragments I and II. One should keep in mind that the position of this new band comprised of the heteroduplexes (see Fig. 2) is variable depending on the electrophoretic conditions as discussed in Section II, B, 2. It is important to run electrophoresis at high temperature for the separation of the open circular DNA from fragments I and II.

2. The existence of the new band is crucial for the next step. If the new

band cannot be detected, one should not proceed to the next step. In this case, the following points should be checked: (a) Was the ratio of fragment I to fragment II approximately 1? (b) Was the mixture maintained at 100°C throughout denaturation? The denaturation can be monitored by gel electrophoresis just following step 2 above. The denatured DNAs should migrate much faster than the double-stranded species. (c) If DNA fragments have a high GC content or if heteroduplexes are not formed for some other reasons, the concentration of the polymerase-ligase buffer may be lowered (add 6 μl instead of 12 μl), or Mg^{2+} can be added after the renaturation step, or one can denature DNA in 1 mM Tris-HCl(pH 7.5) containing 0.1 mM EDTA for 3 min and then add the polymerase-ligase buffer to the DNA solution.

F. Polymerization and Ligation (Step III)

1. Mix the solutions shown in the following tabulation in the order listed (a final volume of 20 μl):

Component	Volume
The remaining reaction mixture from E	11.6 μl
2.5 mM each of dTTP, dCTP, dATP, and dGTP	4 μl
10 mM ATP	2 μl
Klenow enzyme (5 U/μl)	0.4 μl
T4 DNA ligase (0.5 U/μl)	2 μl

2. Incubate the mixture at 12.5°C overnight.

G. Transformation (Step III)

1. Add 1 μl of the reaction mixture described above to 9 μl of H_2O and 2.5 μl 5 × TF and mix with 20 μl of $CaCl_2$-treated competent cells prepared according to Maniatis et al. (1982).

2. Keep the mixture on ice for 30 min followed by incubation in a 42°C water bath for 2 min.

3. Add 0.4 ml of L broth to the mixture and incubate the final mixture at 37°C for another 60 min.

4. Plate 0.1 ml on each of four agar plates containing 50 μg/ml ampicillin (or an appropriate antibiotic).

5. Incubate the plates at 37°C overnight.

5 × TF

>   50 mM CaCl$_2$
>   50 mM MgCl$_2$
>   50 mM Tris-HCl (pH 7.5)

Remarks

The number of transformants per plate should be 200–400.

H. Screening of Transformants

As discussed in Section II,B, there are three methods for screening of desired mutants after transformation. To a certain extent, personal preference may dictate the method used. However, the first method of picking individual transformants (Pick-and-Screen Method), is probably the surest way to detect mutants. It is not particularly labor-intensive to pick only 100 transformants or so, which is enough to isolate mutants if the mutant yields are higher than a few percent. Moreover, this method is recommended if the number of transformants are very few per plate. On the other hand, it one obtains a large number of transformants per plate (more than 200 colonies), the filter methods would be easier. If the mutant yields are less than 1%, the filter methods are more advantagous than the first method. Of the two filter methods described below (Filter Assay Method I and Filter Assay Method II), the method using nitrocellulose filters (Filter Assay Method II) usually gives cleaner results, although it is more expensive than the method using Whatman No. 1 filter paper (Filter Assay Method I). All the agar plates used in this section should contain an appropriate antibiotic according to the plasmid vector used.

1. Pick-and-Screen Method

1. Transfer the individual colonies to a master plate to make streaks of 5–7 mm length.

2. Incubate the plates (master plate) at 37°C for several hours.

3. Transfer cells from the master plates onto the sterile filter using sterile applicator sticks (the ends of which should be smoothed by filing). The filter paper is prepared as follows: Whatman 3MM filter paper is cut to 8 × 8 cm squares on which lines are pencilled every 5 mm to make grids containing a total of 196 small squares for spotting cells. The paper is sterilized by autoclaving, and carefully placed on a prewarmed 100 × 100-mm square plate. (Care should be taken not to trap air bubbles under the filter paper.) Cells should be evenly spotted on the filter by gently turning the applicator sticks on the filter.

4. Incubate the plates face-side up at 37°C overnight.

5. Remove the filter papers from the plates in order to fix the DNA onto the filter paper as described in steps 6–11.

6. Float the filter on 20 ml of 0.5 N NaOH in a glass petri dish (diameter 15 cm) at room temperature for 7 min. Remove the solution by aspiration and add 20 ml of 0.5 N of NaOH gently from the bottom to float the filter paper again. Incubate for another 7 min.

7. Exchange the solution in the petri dish with 30 ml of 0.5 M Tris-HCl (pH 7.4) in the manner described in step 6. The filter is incubated for 3.5 min at room temperature. Repeat.

8. Exchange the solution with 30 ml of 1 × SSC, and gently swirl the petri dish for 3.5 min at room temperature. Repeat.

9. Exchange the solution with 50 ml of 95% ethanol, and vigorously swirl the petri dish for 7 min at room temperature. Repeat.

10. Dry the filter on Whatman 3MM paper for 20–30 min at room temperature.

11. Sandwich the filter paper between Whatman 3MM paper, wrap it in aluminum foil, and bake it for 2 hr at 80°C *in vacuo*.

1 × SSC

150 mM NaCl
15 mM Sodium Citrate (pH 7.2)

2.   Filter Assay Method I

1. Cut out circular Whatman No. 1 Chromatography papers (8 cm in diameter).

2. Put a pencil mark at an appropriate position so that one can determine the proper orientation.

3. Wrap with aluminum foil and sterilize by autoclaving.

4. Place a paper carefully on the agar surface of a plate containing 200–1000 transformants. The size of the colonies are preferably 1–2 mm in diameter. Air bubbles should not be trapped underneath the paper. If the plates have been kept in a refrigerator, they must be warmed before transferring colonies onto the paper. Otherwise the cells are completely transferred to the paper and the colonies cannot be recovered from the master plate.

5. Mark the position of the paper on the petri dish. Gently peel off filters using a pair of blunt-ended forceps.

6. Transfer the paper onto a fresh agar plate keeping the colony-side up and incubate both this and the master plate at 37°C overnight.

7. DNA is then fixed to the filter exactly as described in the previous

section, except for as adjustment of the solution volumes at each step: step 6, 5 ml of 0.5 N NaOH; step 7, 10 ml of 0.5 M Tris-HCl (pH 7.4); step 8, 20 ml of 1 × SSC; and step 9, 30 ml of 95% ethanol.

3. Filter Assay Method II

1. Mark a circular nitrocellulose filter (Schleicher & Schuell; 8.2 cm in diameter) with a waterproof marker pen or ballpoint pen for orientation purposes and autoclave for 3 min (longer autoclaving makes the filter brittle).

2. While colonies are very small (approximately 0.5–1.0 mm in diameter), place the filter on the surface of an agar plate so as to avoid trapping air bubbles trapped underneath the filter and copy markings to the plate.

3. Gently remove the filter using blunt-ended forceps and, with the colony-side up, transfer it to an agar plate containing chloramphenicol (150 μg/ml). Incubate the plate at 37°C overnight. Incubate the original master plate at 37°C for several hours until colonies become visible.

4. Wet a sheet of Whatman 3 MM paper with 0.5 M NaOH/1.5 M NaCl. Remove the excess liquid. (This filter should not be too wet.)

5. Remove replica nitrocellulose filters from plates and place on NaOH/NaCl-saturated filter paper (colony-side up) for 10–15 min at room temperature.

6. Transfer filters to a sheet of Whatman 3MM paper that has been wetted with 1 M Tris-HCl (ph 7.0)/1.5 M NaCl for 2–3 min.

7. Immerse filter in 3× SSC for 15–20 sec.

8. Place on dry Whatman 3MM paper and allow to air-dry.

9. Bake filters at 80°C *in vacuo* for 2 hr.

10. To remove bacterial debris, wash filters again in 3× SSC containing 0.1% SDS using 50–100 ml per filter at 65°C for 1 hr.

11. Change the solution once and incubate for another hour.

12. Wash off any bacterial debris by gently rubbing the filters with fingers (wear gloves).

13. Change solution two more times incubating the filter at 65°C for 1 hr each time.

14. Air-dry on filter paper.

15. Store dry at 4°C or use immediately.

4. Remarks

For Filter Assay Methods I and II, DNA may be fixed directly after transferring colonies from the master plate to the filter without overnight incubation of the filter on an agar plate.

I. Hybridization

1. Put filters in a heat-sealed bag or a zip-lock bag. In the case of nitrocellulose filters, wet the filters with 3× SSC.

2. Use 4 ml of the hybridization solution per 8 × 8-cm Whatman 3MM square filter or 2 ml for a round Whatman No. 1 filter or nitrocellulose filter.

Hybridization Solution

Component	Volume/ml
50 × Denhardt's solution	0.1 ml
30 × NET	0.2 ml
20% Dextran sulfate (Pharmacia)	0.5 ml
10% Sodium dodecyl sulfate	0.05 ml
^{32}P-Labeled oligonucleotide[a] plus H$_2$O	

[a] ^{32}P-labeled oligonucleotide is added at a final concentration of 5×10^5 cpm/ml.

3. Hybridization is carried out for 16–18 hr at an appropriate temperature calculated for each oligonucleotide. The hybridization temperatures are calculated by the following equation empirically derived for synthetic oligonucleotide hybridization (Suggs *et al.*, 1981): [number of A's plus T's × 2 + number of G's plus C's × 4] −4. If the calculated temperature is higher than 55°C, use 50°C for hybridization. *Example:* For a 15-mer oligonucleotide consisting of 7(A + T) and 8(G + C), the hybridization temperature is 42°C = (7 × 2 + 8 × 4) − 4.

4. After hybridization, wash the filters with 100–200 ml of 6× SSC three times for 20 min at room temperature.

5. Wash the filters in 95% ethanol twice for 10 min. In the case of nitrocellulose filters, skip this procedure.

6. Air-dry the filters and place on cardboard. Also fix radioactive position markers on the cardboard.

7. Using an intensifying screen, expose Kodak XAR-5 X-ray film to the filters for 2–5 hr.

8. After developing the film, positive colonies should appear as dark spots (Fig. 4). Orient the filters on the developed film using the radioactive position markers. Transfer the filter markings to the film.

9. Pick positive colonies individually and isolate plasmid DNAs from the 1-ml cultures as described by Birnboim and Doly (1979).

10. Retransform competent cells under dilute DNA conditions with the plasmids and pick 20–30 colonies for each positive isolate.

11. Rescreen the positive colonies using the pick-and-screen method as described above.

50 × Denhardt's Solution (Denhardt, 1966)

Ficoll	1 g
Poly(vinylpyrrolidone)	1 g
BSA (Sigma, Fraction V)	1 g
H_2O	to 100 ml

30 × NET

4.5 *M* NaCl
0.45 *M* Tris-HCl (pH 7.5)
0.03 *M* EDTA

Remarks

1. If washing the filters at room temperature does not give clear positives because of high background (in particular, for those which are done at high hybridization temperatures), filters can be washed at 37°C. If this does not give satisfactory results, wash them at 45°C.

2. If mutations are designed to eliminate or create a restriction endonuclease site, enzyme analysis can be performed instead of the second screening step.

3. No matter which method is used for screening, it is essential to confirm the mutation by DNA sequencing.

IV. REMARKS AND CONCLUSIONS

The techniques for oligonucleotide-directed site-specific mutagenesis using double-stranded plasmid DNA as a template have been dramatically improved. Using the methods outlined in this article, mutant yields can be achieved which are comparable to those obtained with the M13 single-stranded vector. In fact, the use of a double-stranded plasmid as template has various advantages over the M13 method as discussed in Section II. A key step for the method described in this article is heteroduplex formation (step II). The formation of open-circular DNAs can be quantitatively detected by agarose gel electrophoresis of the mixture of fragments I and II after step II. This is an essential check point that determines whether one can proceed to the next step or not.

   The size of the gapped region is determined by the available restriction sites A and B (see Fig. 2). If the size becomes too large, various problems may arise, including nonspecific binding of the oligonucleotide to the single-stranded region. This problem can be easily overcome by adding a DNA fragment(s) to cover a part of the gapped region. Such a fragment can be purified from the plasmid, and added to the denaturation mixture (step II). After the renaturation step, the gap is partially covered by the DNA fragment, and thus one can keep the target site in a small single-stranded region. For example, in the experiment using pIN-III-A3 (Section III,D), the *Hinc*II–*Hin*dIII fragment (0.7 kb; see Fig. 5) can be isolated from the *Pst*I–*Hin*dIII fragment (1.0 kb), and then added back to the denaturation mixture, if the target site is within the *Pst*I–*Hinc*II region (0.3 kb) of the gene for β-lactamase. In this fashion, one can confine the target site to a small region.

   Oligonucleotide-directed site-specific mutagenesis is one of the most powerful techniques in recombinant DNA technology, and will soon be used routinely in molecular biology laboratories for the study of the molecular mechanism of gene regulation, gene structures, functions and structures of proteins, and engineering of new enzymes.

ACKNOWLEDGMENTS

   The authors are grateful to Drs. Martin Teintze, Stephen Pollitt, Pamela J. Green, and Ms. Susan Lehnhardt for their critical reading of this article. The work is supported by Grants GM19043 and GM26843 from the National Institute of General Medical Sciences, Grant NB387I from the American Cancer Society, and a grant from New York State Science and Technology Foundation.

REFERENCES

Barnes, W. M. (1980). *Genet. Eng.* **2,** 185–200.
Birnboim, H. C., and Doly, J (1979). *Nucleic Acids Res.* **7,** 1513–1523.
Charles, A. D., Gautier, A. E., Edge, M. D., and Knowles, J. R. (1982). *J. Biol. Chem.* **257,** 7930–7932.
Cordell, B., Bell, G., Tisher, E., DeNoto, F. M., Ullrich, A., Pictet, R., Rutter, W. J., and Goodman, N. M. (1979). *Cell (Cambridge, Mass.)* **18,** 533–543.
Courtney, M., Jallat, S., Tessier, L.-H., Benavente, A., Crystal, R. G., and Lecocq, J.-P. (1985). *Nature (London)* **313,** 149–151.
Dalbadie-McFarland, G., Cohen, L. W., Riggs, A. D., Morin, C., Itakura, K., and Richards, J. H. (1982). *Proc. Natl. Acad. Sci. U.S.A.* **79,** 6409–6413.
Denhardt, D. T. (1966). *Biochem. Biophys. Res. Commun.* **23,** 641.

Gillam, S., and Smith, M. (1979a). *Gene* **8,** 81–87.

Gillam, S., and Smith, M. (1979b). *Gene* **8,** 99–106.

Gillam, S., Jahnke, D., Astell, R. C., Phillips, S., Hutchinson, C. A., and Smith, M. (1979). *Nucleic Acids Res.* **6,** 2973–2985.

Gillam, S., Astell, R. C., and Smith, M. (1980). *Gene* **12,** 129–137.

Gutteridge, S., Sigal, I., Thomas, B., Arentzen, R., Cordova, A., and Lorimer, G. (1984). *EMBO J.* **3,** 2737–2743.

Hutchinson, C. A., Phillips, S., Edgell, M. H., Gillam, S., Jahnke, P., and Smith, M. (1978). *J. Biol. Chem.* **253,** 6551–6560.

Inouye, S., Soberon, X., Franceschini, T., Nakamura, K., Itakura, K., and Inouye, M. (1982). *Proc. Natl. Acad. Sci. U.S.A.* **79,** 3438–3441.

Inouye, S., Franceschini, T., Sato, M., Itakura, K., and Inouye, M. (1983a). *EMBO J.* **2,** 87–91.

Inouye, S., Hsu, C.P.S., Itakura, K., and Inouye, M. (1983b). *Science* **221,** 59–61.

Inouye, S., Vlasuk, G. P., Hsuing, H., and Inouye, M. (1984). *J. Biol. Chem.* **259,** 3729–3733.

Itakura, K., Rossi, J. J., and Wallace, R. B. (1984). *Annu. Rev. Biochem.* **53,** 323–356.

Kudo, I., Leinweber, M., and RajBhandary, U. L. (1981). *Proc. Natl. Acad. Sci. U.S.A.* **78,** 4753–4757.

Lewis, D. E., Chen, S., Kumar, A., Blanck, G., Pollack, R. D., and Monley, J. L. (1983). *Proc. Natl. Acad. Sci. U.S.A.* **80,** 7065–7069.

Maniatis, T., Fritsch, E. F., and Sambrook, J. (1982). "Molecular Cloning: A Laboratory Manual." Cold Spring Harbor Lab., Cold Spring Harbor, New York.

Masui, Y., Mizuno, T., and Inouye, M. (1984). *Bio/Technology* **2,** 81–85.

Maxam, A., and Gilbert, M. (1980). *In* "Methods in Enzymology" (L. Grossman and K. Moldave, eds.), Vol. 65, pp. 499–560.

Messing, J., Crea, R., and Seeburg, P. H. (1981). *Nucleic Acids Res.* **9,** 309–321.

Montell, C., Fisher, E. F., Caruthers, M. H., and Berk, A. J. (1982). *Nature (London)* **295,** 380–384.

Morinaga, Y., Franceschini, T., Inouye, S., and Inouye, M. (1984). *Bio/Technology* **2,** 636–639.

Oostra, B. A., Harvey, R., Ely, B. K., Markham, A. F., and Smith, A. E. (1983). *Nature (London)* **304,** 456–460.

Perry, L. J., and Wetzel, R. (1984). *Science* **226,** 555-557.

Pielak, G. J., Mauk, A. G., and Smith, M. (1985). *Nature (London)* **313,** 152–154.

Razin, A., Hirose, T., Itakura, K., and Riggs, A. D. (1978). *Proc. Natl. Acad. Sci. U.S.A.* **75,** 4268–4270.

Seeburg, P. H., Colby, W. W., Capon, D. J., Goeddel, D. V., and Levinson, A. D. (1984). *Nature (London)* **312,** 71–75.

Simons, G. F. M., Veeneman, G. H., Konings, R. N. H., Van Boom, J. H., and Schoenmakers, J. G. G. (1982). *Nucleic Acids Res.* **10,** 821–832.

Suggs, S., Wallace, R. B., Hirose, T., Kawashima, E., and Itakura, K. (1981) *Proc. Natl. Acad. Sci. U.S.A.* **78,** 6613–6617.

Villafranca, J. E., Howell, E. E., Voet, D. H., Strobel, M. S., Ogden, R. C., Abelson, J. N., and Kraut, J. (1983). *Science* **222,** 782–788.

Vlasuk, G. F. M., and Inouye, S. (1983). *In* "Experimental Manipulation of Gene Expression" (M. Inouye, ed.), pp. 291–303. Academic Press, New York.

Vlasuk, G. P., Inouye, S., Ito, H. Itakura, K., and Inouye, M. (1983). *J. Biol. Chem.* **258,** 7141–7148.

Vlasuk, G. P., Inouye, S., and Inouye, M. (1984). *J. Biol. Chem.* **259,** 6195–6200.

Wallace, R. B., Johnson, P. F., Tanaka, S., Schold, M., Itakura, K., and Abelson, J. (1980). *Science* **209**, 1396–1400.

Wallace, R. B., Johnson, M. J., Suggs, S. V., Miyoshi, K., Blatt, R., and Itakura, K. (1981a). *Gene* **16**, 21–26.

Wallace, R. B., Schold, M., Johnson, M. J., Dembek, P., and Itakura, K. (1981b). *Nucleic Acids Res.* **9**, 3647–3656.

Wang, A., Lu, S.-D., and Mark, D. F. (1984). *Science* **224**, 1431–1433.

Wilkinson, A. J., Fersht, A. R., Blow, D. M., Carter, P., and Winter, G. (1984). *Nature (London)* **307**, 187–188.

Winter, G., Fersht, A. R., Wilkinson, A. J., Zoller, M., and Smith, M. (1982). *Nature (London)* **299**, 756–758.

Zoller, M. J., and Smith, M. (1982). *Nucleic Acids Res.* **10**, 6487–6500.

Zoller, M. J., and Smith, M. (1983). *In* "Methods in Enzymology" (R. Wu, L. Grossman, and K. Moldave, eds.), Vol. 100, pp. 468–500.

9

Synthetic Oligonucleotides for the Identification and Isolation of Specific Gene Sequences

C. GARRETT MIYADA
ANNA B. STUDENCKI
R. BRUCE WALLACE
Department of Molecular Genetics
Beckman Research Institute of the City of Hope
Duarte, California 91010

I. INTRODUCTION

Watson and Crick (1953) first elucidated the internucleotide hydrogen-bonding rules responsible for the stability of duplex DNA. Hydrogen bonding of adenine with thymine or of cytosine with guanine is referred to as Watson–Crick base pairing. Watson–Crick base pairing is the foundation for the specificity in hybridizations of synthetic oligodeoxyribonucleotides (henceforth referred to as oligonucleotides) to complementary DNA or RNA molecules. Intuitively, one could deduce that mispaired bases should disrupt the stability of the hydrogen-bonded oligonucleotides; however, it was not until the two separate techniques of DNA sequence determination and the chemical synthesis of short oligomers of DNA were combined that any quantitative data on the effects of base pair mismatches on duplex stability could be obtained. Under appropriate hybridization conditions it was shown that oligonucleotides required complete complementarity with its cognate sequence for duplex formation.

Synthesis and Applications
of DNA and RNA

Thus, oligonucleotide hybridization probes have the ability to distinguish RNA or DNA sequences that differ by as little as a single base (pair). It is this specificity that has led to the development of synthetic oligonucleotides as a standard tool for molecular biologists in the discrimination of closely related gene sequences.

Synthetic oligonucleotides have been incorporated in many molecular biology protocols. Oligonucleotides have been used to synthesize genes (Itakura et al., 1977; Crea et al., 1978), to introduce restriction enzyme sites (linkers, Scheller et al., 1977), to introduce specific base pair alterations in a defined gene (Hutchison et al., 1978; Razin et al., 1978; Wallace et al., 1980), and to prime DNA synthesis from either a DNA or an RNA template for the purpose of primary sequence determination. This chapter will deal primarily with the use of oligonucleotides for the identification and isolation of specific gene sequences. It will also cover the use of oligonucleotides in the detection of human genetic diseases. For other recent reviews that cover similar subject matter, the reader is referred to the following (Smith, 1983; Itakura et al., 1984; Wallace and Itakura, 1984; Reyes and Wallace, 1984).

II. EARLY STUDIES

Experiments from several laboratories have been important in the development of synthetic oligonucleotides as hybridization probes for gene sequences. The thermal stability of base-paired oligonucleotides was shown to increase with the length of the oligonucleotide (Niyogi and Thomas, 1968; Astell and Smith, 1972). Specificity of oligonucleotide binding to DNA was demonstrated with hybridization experiments to the cohesive ends of phage lambda (Wu, 1972) or to one strand of strand-separated phage DNAs (Besmer et al., 1972; Doel and Smith, 1973). Oligonucleotides were also shown to bind specifically to an mRNA species (Doel and Smith, 1973). Precise sequence specificity was demonstrated in experiments using an oligonucleotide as a primer for DNA polymerization off either a DNA (Wu et al., 1973) or an RNA (Szostak et al., 1977) template. Single base pair mismatches were shown to reduce greatly the stability of a base-paired oligonucleotide (Gillam et al., 1975; Wallace et al., 1979). Finally, the 5'-OH end of an oligonucleotide can be labeled to a very high specific activity with polynucleotide kinase and [γ-^{32}P]ATP, which makes them useful as probes in Southern (Southern, 1975), colony (Grunstein and Hogness, 1975), and plaque (Benton and Davis, 1977) hybridizations.

III. IDENTIFICATION OF CLONED DNA SEQUENCES BASED ON THE PRIMARY PROTEIN SEQUENCE

There are generally two methodologies used to isolate the DNA-encoding sequences for an isolated protein. In one, the isolated protein is used to raise antibodies against it. Complementary DNA (cDNA) is prepared from mRNA isolated from cells producing the protein of interest. The cDNA is then cloned in a suitable vector, with the hope of expressing a chimeric protein that possesses some of the same antigenic determinants found in the isolated protein. Antibody screening can then be used to detect clones potentially containing the desired cDNA sequence (Villa-Komaroff *et al.*, 1978). Alternatively, the protein can be sequenced and the primary amino acid sequence used to design appropriate oligonucleotide probes (Wu *et al.*, 1973; Agarwal *et al.*, 1981; Wallace *et al.*, 1981). Since only two amino acids are encoded by a unique DNA sequence, there is a redundancy in the potential DNA sequences that may encode a given protein sequence. Overcoming this redundancy has led to several different strategies in the design of oligonucleotides complementary to DNA or RNA sequences encoding a given protein sequence. These strategies will be reviewed in this section.

A. Unique Primers and Probes

When a given protein sequence shows little redundancy with respect to its potential coding DNA sequences, a unique oligonucleotide can be used to identify the cloned gene sequence of interest. Such an oligonucleotide is designed with potential $G:U$ or $G:T$ base pair mismatches whenever the ambiguous base is either one of two purines or one of two pyrimidines (Wu *et al.*, 1973; Agarwal *et al.*, 1981). It has been shown that $G:T$ and $G:U$ base pair mismatches are much less destabilizing to oligonucleotide hybridization than the alternative purine:pyrimidine mismatch, $A:C$ (Agarwal *et al.*, 1981). The oligonucleotide is then either characterized by its ability to prime unique cDNA transcripts (Cheng *et al.*, 1976; Szostak *et al.*, 1977; Rabbitts *et al.*, 1977) or its specificity in a Northern (Alwine *et al.*, 1977) hybridization. This strategy was used to identify cDNAs for the hog gastrin gene (Noyes *et al.*, 1979), rat insulin genes I and II (Chan *et al.*, 1979), rat relaxin gene (Hudson *et al.*, 1981), bovine (Gubler *et al.*, 1981) and human (Legon *et al.*, 1982) preproenkephalin genes, and bacteriorhodopsin gene (Chang *et al.*, 1981). In most of the cases (Noyes *et al.*, 1979; Chan *et al.*, 1979; Gubler *et al.*, 1981; Legon *et al.*, 1982) the specificity of the priming reaction was verified by sequencing unique cDNA transcripts and identifying the appropriate codons upstream of the priming site. Either oligonucleotide-primed cDNA (Chan *et al.*, 1979; Hudson *et al.*, 1981; Chang *et al.*, 1981) or the oligonucleotide itself (Gubler *et al.*, 1981; Legon *et al.*, 1982) was then used to screen cDNA libraries in the identification of the desired gene sequence.

An oligonucleotide primer based on a protein sequence sometimes generates many cDNA transcripts, especially if the correctly hybridizing mRNA species is rare. In this case, the appropriate cDNA transcript can still be identified by priming cDNA synthesis in the presence of one dideoxy-NTP (ddNTP) and the remaining three dNTPs (Sood *et al.*, 1981; Stetler *et al.*, 1982). This results in a staggering of cDNA transcripts that are separated according to length on a denaturing polyacrylamide gel. The longer a transcript is in this reaction, the greater the chance that it is due to a unique priming event. Knowledge of the amino acid sequence upstream from the priming site aids in the determining which ddNTPs to use and in identifying the proper cDNA transcript by DNA sequence analysis. Once the correct cDNA transcript is identified, the additional DNA sequence information can be used to design a more specific (longer) oligonucleotide probe. This strategy was used to identify the human histocompatibility *HLA-B7* gene (Sood *et al.*, 1981) and the human *HLA-DR* α-chain gene (Stetler *et al.*, 1982) cDNAs.

In the previous studies, in which unique oligonucleotides were used to prime specific cDNAs, the protein sequence used to design the oligonucleotide was generally from the amino terminus. Oligonucleotides specific for the 5' end of a transcript are usually better primers for the identification of specific cDNA since they are more likely to generate a unique runoff transcription product. Oligonucleotide primers specific for the 3' end of genes or their transcripts have also been useful in cloning since they can be used to generate primer-enriched cDNA libraries. While most cDNA libraries are primed in a nonspecific manner with oligo(dT), primers of the sequence $d(pT_8-N-N')$ have been shown to prime cDNA synthesis in a sequence-specific manner adjacent to the poly(A) tract for a given mRNA (Cheng *et al.*, 1976; Sasavage *et al.*, 1980). The oligonucleotide $d(pT_8-G-C)$ was used to enrich for insulin cDNAs in the construction of a β-lactamase-insulin protein fusion (Villa-Komaroff *et al.*, 1978). Primers specific for the 3'-end of the genome have also been useful in the cloning of genomic cDNA copies of the negative-strand RNA virus, influenza (Lai *et al.*, 1980; Winter *et al.*, 1981).

B. Mixed Oligonucleotide Sequences

In general, it is not possible to design a unique oligonucleotide for a given protein sequence. For example, the most common amino acids found in a study of human gene sequences were leucine and serine, both of which are encoded by six possible codons (Lathe, 1985). The same study found that methionine and tryptophan, the only amino acids encoded by unique sequences, are the most infrequently used amino acids. One strategy in designing hybridization probes is to synthesize a mixture of oligonucleotides that will include all possible coding sequences for a given peptide sequence (Wallace *et al.*, 1981). The mixture of oligonucleotides can then be used in a similar manner as the previously described unique oligonucleotide sequences for cDNA cloning. Mixed oligonucleotides can be used directly as hybridization probes, as primers for cDNA hybridization probes, and as primers to enrich for certain sequences in the construction of a cDNA library.

Model hybridization studies for a mixture of oligonucleotide sequences based on a single protein sequence were originally performed by Wallace and co-workers (1981). They found that precise hybridization of a unique 13-base sequence in a mixture of seven related 13-mers could be obtained in both Southern and colony hybridization protocols. The experimentally derived T_m of the oligonucleotide mixture was not substantially different from the T_m derived from a unique complementary sequence.

Subsequently mixed oligonucleotide probes were used for the identification of cDNA clones for the human β_2-microglobulin gene (Suggs *et al.*, 1981) and a murine H-$2K^b$-related molecule (Reyes *et al.*, 1981). The first 1982) and human β subunit of nerve growth factor (Ullrich *et al.*, 1983). and another mixture of sixteen 15-mers, with both sets based on the same protein sequence. The second study utilized a set of eight 16-mers. Both studies also utilized a set of eight 11-mers based on a sequence from a different region of the protein. One 11-mer mix was successful in identifying the proper cDNA clone in Southern hybridizations but did not detect the same clone in a colony hybridization (Suggs *et al.*, 1981). Mixtures as complex as 384 23-mers have been successful in the identification of cDNA clones in colony hybridization procedures (Whitehead *et al.*, 1983). Another useful technique in identifying the cDNA clone of interest is to hybridize with two sets of oligonucleotides, with each set based on a different peptide sequence (Suggs *et al.*, 1981; Reyes *et al.*, 1981; Woods *et al.*, 1982; Noda *et al.*, 1982). This double-screening procedure eliminates some of the false positives obtained from the initial screening.

Mixed oligonucleotides have also been used to prime the synthesis of radioactive cDNA, which was later used as a hybridization probe in the screening of a cDNA library. This approach proved successful in the detection of cDNA clones for human leukocyte interferon (Goeddel *et al.*, 1980a), human fibroblast interferon (Goeddel *et al.*, 1980b), and rat angiotensinogen (Ohkubo *et al.*, 1983). In all of these examples the oligonu-

cleotide-primed cDNAs were only used for the initial screening. In one
case the original positives were rescreened with a mixed oligonucleotide
probe (Goeddel et al., 1980a). In another case a specific transcript (isolated from the original cDNA priming and purified by denaturing polyacrylamide gel electrophoresis) was used in the final screening (Ohkubo et
al., 1983).

Another strategy has been to use a set of mixed oligonucleotides to
prime cDNA synthesis in the construction of a library. The primer-enriched cDNA library is then screened with a mixed oligonucleotide probe
based on a peptide sequence closer to the amino terminus of the protein
than the original priming sequences. This technique was successful in
obtaining cDNA clones for human antihemophilic factor IX (Choo et al.,
1982) and human β subunit of nerve growth factor (Ullrich et al., 1983).

Finally, a recent development by Wood and colleagues (1985) should
aid in the determination of the discriminatory wash temperature required
to let only the exactly complementary oligonucleotide in a mixture hybridize. They have used the salt tetramethylammonium chloride to replace the standard saline citrate solution in discriminatory washes following oligonucleotide hybridization. Tetramethylammonium chloride
eliminates the preferential melting of A:T versus G:C base pairs,
thereby creating a linear relationship between the T_m of a mixed oligonucleotide probe and its length in bases. This protocol has proved successful
in identifying cDNA clones for the γ-subunit of mouse nerve growth
factor among a set of clones encoding related kallikrein gene sequences
(Ullrich et al., 1984a).

C. Long Unique Oligonucleotide Probes

While a mixture of oligonucleotides has been successful in the isolation
of cloned gene sequences, the methodology has some drawbacks as the
complexity of the probe mixture increases. As the number of sequences
increases, there is a greater chance that the probe will hybridize fortui-
tously to a random cDNA sequence. In cases where the desired gene
sequence is produced as a rare mRNA, the problems in detecting the
clone of interest become compounded. As an alternative to mixed oligonucleotides, long unique oligonucleotides have been developed as hybridization probes for gene sequences. In this approach the oligonucleotide sequence is based on codon usage frequencies (Grantham et al.,
1980) and probe length confers its specificity rather than absolute complementarity (Lathe, 1985).

Perhaps the earliest example of ignoring codon degeneracy in the design of an oligonucleotide probe was described in the identification of the

5' mRNA sequence of human fibroblast interferon (Houghton *et al.*, 1980). A pentapeptide sequence, which could be encoded by 512 oligonucleotide sequences with seven ambiguous positions, was compared to homologous amino acid sequences from mouse interferons. The authors' analysis led to the design of two unique 15-mer sequences. Both sequences were checked for specificity as primers in a cDNA synthesis reaction. One oligonucleotide demonstrated a probable homology to interferon mRNA, since it primed a unique transcript only with RNA isolated from induced fibroblasts. The primer-specific cDNA was isolated, and its nucleotide sequence was determined and was found to encode the 5' end of the fibroblast interferon mRNA.

Jaye *et al.* (1983) described the first successful use of a long oligonucleotide to screen a cDNA library. In this case they identified a clone containing bovine antihemophilic factor IX sequences with a 52-base oligonucleotide. The same strategy has been successful in identifying cDNA clones for the human epidermal growth factor gene (Ullrich *et al.*, 1984b), the human tumor necrosis factor gene (Pennica *et al.*, 1984), and the human insulin receptor gene (Ullrich *et al.*, 1985). In the previously mentioned studies, the oligonucleotide length varied from 42 bases (Pennica *et al.*, 1984) to 63 bases (Ullrich *et al.*, 1985). The probe homology to the actual coding sequence varied from 78% (Ullrich *et al.*, 1985) to 90% (Ullrich *et al.*, 1984b). The probe homologies could actually have been better but in all of the cases the protein sequence used in designing the oligonucleotide contained either one or two incorrect amino acids.

Long oligonucleotides have also been successful in the identification of genomic clones. In these cases cDNA cloning had been either unsuccessful (Derynck *et al.*, 1984) or unfeasible because the site of protein synthesis was unknown (Anderson and Kingston, 1983; Wood *et al.*, 1984). A bovine pancreatic trypsin inhibitor clone was identified with an 86-base oligonucleotide (Anderson and Kingston, 1983), a human transforming growth factor-α clone with a 74-base probe (Derynck *et al.*, 1984), and a human factor VIII clone with a 36-base probe (Wood *et al.*, 1984). In one case, a 48-base oligonucleotide (whose sequence is included in the successful 74-base probe) did not hybridize specifically to the appropriate genomic clone even though it shared a 77% homology with it (Derynck *et al.*, 1984). In all of the other studies the probe shared at least a 77% homology with its genomic counterpart. In two of the studies the genomic DNA sequence was used to clone its related cDNA sequence (Derynck *et al.*, 1984; Gitschier *et al.*, 1984). These results suggest that the cloning of extremely rare mRNA species may first require the isolation of its genomic counterpart.

While all of the previously mentioned studies made use of mammalian preferential codon usage, long oligonucleotide probes exhibiting an *E. coli* codon usage bias have been successful in identifying human gene sequences. Human insulin-like growth factor I gene sequences (Ullrich *et al.*, 1984c) as well as human lymphotoxin gene sequences (Gray *et al.*, 1984) were detected by using sequences originally designed for high protein expression from an *E. coli* plasmid vector.

Lathe (1985) has recently tabulated a set of rules for designing long unique oligonucleotide probes based on human gene sequences. In addition to determining optimal codons for each amino acid, he also analyzed factors that should be considered in determining the length for a given oligonucleotide probe. For 85% probe homology (a result that should be obtained with an average human protein sequence), Lathe suggests a probe length of 36 bases for screening a cDNA library and a probe of 62 bases for screening a genomic library.

IV. DETECTION OF SPECIFIC RNAs

The gene specificity of oligonucleotides to RNAs was originally demonstrated in liquid hybridizations to RNA isolated from either a wild-type organism or a mutant deleted for the gene of interest. A 12-base oligonucleotide was shown to be specific for the lysozyme gene mRNA of bacteriophage T4 (Doel and Smith, 1973). In this study 7-base and 9-base oligonucleotides did not hybridize in a gene specific manner. A 15-base oligonucleotide was found to be specific for mRNA encoded by the yeast iso-1-cytochrome *c (CYC-1)* gene (Szostak *et al.*, 1977). In both of these studies, a unique oligonucleotide sequence was deduced based on the amino acid sequence of wild-type and frameshift mutant proteins. With the availability of the gastrin cDNA sequence, 8-mer, 12-mer, and 15-mer oligonucleotides were designed to quantitate gastrin mRNA (Mevarech *et al.*, 1979). The 8-mer detected two bands in Northern hybridizations while the 12-mer and 15-mer were both specific for gastrin mRNA. The 12-base probe was also shown to detect gastrin mRNA in a quantitative manner in both Northern and dot blot hybridizations.

The base-pair specificity of oligonucleotides in hybridizations to mRNA was demonstrated with probes designed to detect human preproenkephalin mRNA (Legon *et al.*, 1982). Two 14-base probes, whose sequences encoded the same primary protein sequence but differed by a single base, were tested in Northern hybridizations for their ability to detect preproenkephalin mRNA. Only one probe hybridized to preproenkephalin mRNA and it was later shown that its sequence was completely complementary to its cognate mRNA. The other oligonucleotide contained an A : C mismatch with respect to the same mRNA sequence. This study demonstrated the highly discriminatory nature of oligonucleotide hybridization to RNAs and suggested that oligonucleotides could be used to differentiate transcription from related genes. Rat cytochrome *P-450b* and *P-450e* mRNAs were individually quantitated in liver RNA preparations by using 18-base-long oligonucleotides (Omiecinski *et al.*, 1985). In this example, the related probe sequences differed at four positions. Transcription specific for the murine transplantation antigen gene, *H-2K^b*, was detected with a 19-base oligonucleotide in cells expressing other related transplantation antigens (Reyes and Wallace, 1984). Tissue-specific transcription of another murine major histocompatibility complex (*H-2*) class I gene, *Q10,* was detected with a 23-base oligonucleotide (Miyada *et al.*, 1985).

V. DETECTION OF SPECIFIC CHROMOSOMAL SEQUENCES

Amino acid sequencing of frameshift mutant proteins led to the unambiguous determination of 44 consecutive nucleotides in the 5'-end coding sequence of the yeast *CYC-1* gene (Stewart and Sherman, 1974). A 13-base oligonucleotide based on a portion of the above nucleotide sequence was used to probe a yeast genomic restriction digest to identify the *CYC-1* gene (Montgomery *et al.*, 1978). The probe detected three strongly hybridizing restriction fragments even though, statistically, a 13-base sequence should be unique within the yeast genome (Astell and Smith, 1972). Hybridizations to restriction digests of *CYC-1* mutant DNAs led to the identification of the hybridizing restriction fragment for the *CYC-1* gene (Montgomery *et al.*, 1978). The 13-mer was then used to detect the *CYC-1* gene in a partial phage gene library. A 15-base oligonucleotide, which overlapped the above 13-base sequence by 5 bases, was shown to detect two restriction fragments in a similar yeast chromosomal restriction digest hybridization (Szostak *et al.*, 1979). One of the hybridizing bands corresponded to the *CYC-1* gene.

The allelic specificity of oligonucleotide hybridization probes in genomic restriction digests was demonstrated in experiments to determine carriers of the genetic disorder sickle cell anemia (Conner *et al.*, 1983). Nineteen-base oligonucleotides were used to discriminate the wild-type β-globin allele (β^A) from the mutant β^S allele, which is responsible for sickle cell anemia in the homozygous state. These alleles differ by a single base pair; however, oligonucleotide hybridization detected either gene in either homozygous or hemizygous states (see Section VI for a more detailed account).

Oligonucleotide probes are generally used to discriminate highly homologous gene sequences, for example, members of a multigene family. However, they may also be used to detect highly conserved sequences common to a gene family and thus identify related genes. This strategy has been used successfully to identify chloroplast tRNA genes in higher plants (Nickoloff and Hallick, 1982) as well as cDNA clones for members of the murine kallikrein gene family (Ullrich *et al.*, 1984a). We have designed an oligonucleotide to a highly conserved region among *H-2* class I genes (Miyada *et al.*, 1985). This probe was used in hybridizations to an *H-2* class I cosmid library (Steinmetz *et al.*, 1982) and was found to hybridize to 32 of 35 restriction fragments (genes) that were detected by a homologous cDNA probe (C. G. Miyada and R. B. Wallace, unpublished results).

Oligonucleotides hybridization probes also appear to be extremely useful in determining complex mechanisms for altering gene structure and expression. Our laboratory has been studying the *bm*1 mutation of the murine transplantation antigen gene, *H-2K^b*. The *bm*1 mutation is the result of seven base-pair changes over a 13-base-pair-region. The clustering of nucleotide substitutions suggested that gene conversion was a likely mechanism for generating this mutation; however, at the time a potential "donor" sequence for the *bm*1 mutation was unknown in the parental genome. By using oligonucleotides as hybridization probes to genomic restriction digests, we demonstrated that there exists a potential donor sequence and this sequence is duplicated in the *bm*1 mutant genome (Miyada *et al.*, 1985). Also the wild-type *H-2K^b* sequence that overlaps the *bm*1 mutation is apparently lost in the mutant genome. Thus, it appears that gene-conversion-like events may be responsible in part for the high diversity found among different alleles of the *H-2* transplantation antigen genes.

The variant surface glycoprotein (VSG) found in *Trypanosoma brucei* is an example of a gene system whose rapid evolution is required to escape immune detection in its host. A 22-base oligonucleotide has been used to detect a 35-base "mini-exon" common to many if not all mRNAs (De Lange *et al.*, 1983). Adjacent to the "mini-exon" sequences are the VSG coding sequences on VSG mRNA. In hybridizations to whole chromosomes, which had been separated by pulsed-field gradient gel electrophoresis, it was shown that the "mini-exon" sequences and the VSG coding sequences reside on different chromosomes (Guyaux *et al.*, 1985). This result implies that transcription products from two different chromosomes are somehow spliced to form a functional mRNA. This mechanism is probably also related to the trypanosomes' capability to readily activate silent VSG genes and express new VSGs with different antigenic determinants.

VI. DETECTION OF HUMAN GENETIC DISEASES

The hemoglobinopathies constitute a group of phenotypically related diseases caused by what are now known to be a vast number of underlying molecular defects in the α- and β-globin gene clusters (for example, see Kazazian, 1985). These defects include deletions and insertions as well as point mutations. Similar diversities in etiology have been discovered for hypoxanthine-guanine phosphoribosyltransferase (HPRT) deficiency and may also be expected for other groups of phenotypically related genetic diseases. When a molecular basis for the genetic defect has been established, diagnosis of the carrier state in adults as well as the disease state *in utero* is possible.

The history of the study of the hemoglobinopathies illustrates how the detection of genetic disease *in utero* has become more and more sophisticated. Initially fetoscopy or placental aspiration was used to obtain fetal erythrocytes, which were then analyzed for the globin proteins synthesized. Both methods carried an unacceptably high risk to the fetus (5–9% mortality rate; Phillips *et al.*, 1980) and have been supplanted by amniocentesis (<0.5% mortality rate; Phillips *et al.*, 1980). This much safer procedure is used to obtain fetal cells from which DNA is prepared for genotyping. Initially solution hybridization of these amniocyte DNAs with radiolabeled cDNAs and, later, Southern blot analysis permitted detection of only those hemoglobinopathies caused by large deletion mutations, e.g., homozygous α^0-thalassemias (hydrops fetalis, Kan *et al.*, 1976; Orkin *et al.*, 1978), certain α^+-thalassemias (Kan *et al.*, 1979; Phillips *et al.*, 1979a; Orkin *et al.*, 1979a), $\delta\beta$-thalassemias (Ottolenghi *et al.*, 1976; Orkin *et al.*, 1978; Mears *et al.*, 1978), and one β^0-thalassemia (Flavell *et al.*, 1979; Orkin *et al.*, 1979b).

Hemoglobinopathies caused by point mutations have yielded more slowly to analysis. More sophisticated methodologies—restriction fragment length polymorphism (RFLP) analysis, direct restriction enzyme site analysis, and now allele-specific hybridization with oligonucleotide probes—had to be devised to distinguish the normal from affected genes. The experimental protocols comprising these three methodologies for detecting point mutations share common elements: Each relies on digestion of DNA samples with restriction enzymes, on electrophoretic separation of the generated restriction fragments, and on the use of radiolabeled DNAs for probing the resolved digests. However, because each approach relies on a different principle for its detection of point mutations, each differs in the degree to which it is applicable.

The first of these methodologies is indirect and depends on the association of a given point mutation with an RFLP. If no such association can be found, this approach is not applicable. Because the frequency of linkage of the point mutation to a particular RFLP is never 100%, this approach requires a family study to determine the pattern of gene segregation among the polymorphic restriction fragments. Unfortunately, unless a second RFLP is found, the genotyping of offspring is not possible in families where both the normal and abnormal genes happen to be associated with the same fragment length in the parents (Phillips et al., 1980). Some point mutations known to be in linkage disequilibrium in some populations for a particular fragment size at an RFLP site include: sickle cell anemia, which is associated with a HpaI polymorphism (Kan and Dozy, 1978, 1980a,b; Phillips et al., 1979b; Panny et al., 1981) and also a HindIII polymorphism (Phillips et al., 1980; Antonarakis et al., 1982); $\beta^{39(UAG)}$-thalassemia, which is associated with a BamHI polymorphism (Kan et al., 1980); and β^C-disease, which is associated with a HpaI polymorphism (Kan and Dozy, 1980b).

The second of these methodologies is dependent on the nature of the point mutation itself. If a point mutation creates or destroys a restriction site, genomic DNAs can be examined directly after digestion with the enzyme specific for that site, followed by Southern blot hybridization analysis. The loss or gain of a restriction site will result in a different banding pattern between the mutant and wild-type DNAs in the resulting autoradiogram. Examples of point mutations which are amenable to direct assay in this fashion are: $\beta^{121\ Glu \rightarrow Lys}$ (Hb O^{Arab}), which is associated with the loss of an EcoRI site (Phillips et al., 1979b); sickle cell anemia with the loss of DdeI (Geever et al., 1981; Chang and Kan, 1981) and MstII (Chang and Kan, 1982; Orkin et al., 1982; Goossens et al., 1983) sites; HPRT$_{Toronto}$ with the loss of a TaqI site (Wilson et al., 1983a,b); β^{IVS} $^{2-1}$ with the creation of an HphI site (Baird et al., 1981); and Hb M Milwaukee with the creation of an SstI site (Oehme et al., 1983). This approach is much simpler than the first because it does not require extensive family studies. However, it is also limited in its applicability since not every point mutation creates or destroys a restriction site.

The newest of the methodologies for detecting point mutations is more generally applicable than the others. Unlike the first two methodologies, which employ nick-translated DNAs as hybridization probes, this one employs synthetic oligonucleotides for this purpose. To insure that oligomer probes are of sufficiently high complexity to be specific only for the genes of interest, a minimum length of 19 nucleotides is chosen for the probes. For each mutation site a set of probes, which includes one oligomer complementary to the normal gene sequence and one complementary to the mutated sequence, is synthesized. Under appropriately stringent hybridization and wash conditions, perfectly matched and mismatched duplexes can be distinguished by the presence or absence of radioactive signal from the particular restriction fragment that carries these genes. Examples of mutant genes which have been distinguished from their normal counterparts by this approach are sickle cell anemia (β^S-globin, Conner et al., 1983; Studencki and Wallace, 1984), β^C-globin (Studencki et al., 1985), α_1-antitrypsin (Kidd et al., 1983, 1984), $\beta^{39(UAG)}$ thalassemia (Pirastu et al., 1983, 1984a,b), and $\beta^{IVS\,1}$-thalassemia genes (Orkin et al., 1983).

It should be pointed out that RFLP linkage analysis has one significant advantage over the other two described methodologies. While the primary structure of the gene and the molecular nature of the defect need to be known for detection of point mutations by direct restriction site analysis or synthetic oligonucleotide hybridization, in RFLP linkage analysis the defective gene need not be known. Both Huntington's disease (Gusella et al., 1983) and Duchenne muscular dystrophy (Davies et al., 1983) have been associated with RFLPs even though the defective genes in both cases have yet to be identified. It has been postulated that approximately 150 RFLPs would be required to map the entire human genome (Botstein et al., 1980). As the number of characterized RFLPs increases, RFLP linkage analysis will become an important initial step in the characterization of inherited diseases.

Detection of Point Mutations of the β-Globin Gene: An Example

The mutant β^S and β^C alleles of the β-globin gene are responsible for a pair of related hemoglobinopathies. The homozygous $\beta^S\beta^S$ state leads to sickle cell anemia while the heterozygous $\beta^S\beta^C$ genotype causes a milder form of anemia. The β^S and β^C alleles differ from the wild type β^A by a single base pair, which both result in an amino acid substitution in codon six. The β^S allele is the result of an A → T transversion in the second codon position leading to a valine for glutamic acid substition. The β^C allele is the result of a G → A transition in the first-codon position causing a lysine substitution for glutamic acid. (The wild-type and mutant β-globin alleles are shown in Table I.)

An illustration of the oligonucleotide hybridization assay for the discrimination of the β^A, β^S, and β^C alleles in human DNA samples is shown in Fig. 1. In this figure human genomic DNA samples were digested with BamHI, subjected to electrophoresis on a 1% agarose gel, and hybridized with 5'-end labeled, 19-base oligonucleotides specific for either the β^A (probe Hβl9A'), β^S (probe Hβ19S), or β^C (probe Hβ19C) alleles. As seen in Fig. 1, accurate genotyping of human DNA is performed by observing

TABLE I

Coding Sequences for the β^A, β^S, and β^C
Alleles of the Human β-Globin Gene

	5	6	7
	Pro	Glu	Glu
β^A	CCT	GAG	GAG[a]
	Pro	Val	Glu
β^S	CCT	GTG	GAG
	Pro	Lys	Glu
β^C	CCT	AAG	GAG

[a] The *Mst*II restriction site (CCTNAGG)
present in the β^A allele DNA sequence is under-
lined.

hybridization of the oligonucleotides to a 1.8-kb restriction fragment.
Homozygous and heterozygous states are easily distinguished (compare
lanes for samples 015 and 016).

For these particular alleles, oligonucleotide hybridization provides a
simpler and more accurate means of genotyping than RFLP linkage analysis or the analysis of restriction site changes. In the Black population the
β-globin gene is associated with a *Hpa*I polymorphism that produces
either 7.0, 7.6, or 13.0 kb restriction fragments. However, neither the
normal β^A globin gene nor either of the two variant genes β^S and β^C is
exclusively associated with any one of these fragment lengths. The β^A
gene is most often linked with the 7.6-kb fragment (88% frequency) while
both the β^S and β^C genes are usually linked with the 13.0-kb fragment, the
β^S gene at a frequency of 68%, and the β^C gene at a frequency of >95%
(Phillips *et al.*, 1980; Chang and Kan, 1981). The β^S allele can be detected
through restriction enzyme analysis since the A $\rightarrow$ T transversion results
in the loss of an *Mst*II site (Chang and Kan, 1982; Orkin *et al.*, 1982;
Goossens *et al.*, 1983) (Table I). However, this type of analysis cannot
distinguish the β^A and β^C alleles for in this case, the *Mst*II site is left
intact.

The applicability of synthetic oligonucleotide hybridization to detect
human genetic diseases is dependent on the number of possible mutant
genotypes. In the case of $\beta^{39(UAG)}$ thalassemia, which is responsible for
over 95% of the β-thalassemia lesions in Sardinia, synthetic oligonucleotide hybridization proved to be very accurate in the identifying the
carrier state as well as genotyping unborn fetuses (91% accuracy in prenatal diagnosis) among Sardinians (Rosatelli *et al.*, 1985). In another β-
thalassemia study, in which Northern Italians and Cypriots were sam-

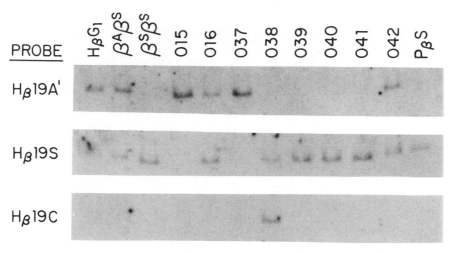

Fig. 1. Direct gel hybridizations of human genomic DNAs with the 5' end-labeled non-adecamer probes Hβ19A', and Hβ19S, Hβ19C (Studencki *et al.*, 1985). Ten-microgram samples of human genomic DNAs from 10 different individuals (two of known genotype, lanes $\beta^A\beta^S$ and $\beta^S\beta^S$; the remaining samples were unknown) were digested with *Bam*HI and were resolved on a large vertical 1% agarose gel (0.3 × 20 × 20 cm; 1-cm wells). *Bam*HI digests of 150 pg of γHβG1 and of 30 pg of pBR322-HβS markers were also included. Electrophoresis was carried out at 40 V for 16 hr in 100 mM Tris-borate (pH 8.3), 1 mM EDTA. The gel was hybridized sequentially with the three probes. Hybridization mixes contained either 3.5 × 10^6 cpm (approx. 2 ng) Hβ19A', 2.7 × 10^6 cpm Hβ19S or 5.8 × 10^6 cpm Hβ19C probe per milliliter in 0.9 M NaCl, 0.18 M Tris-HCl (pH 8.0), 6 mM EDTA, 0.5% Nonidet P-40, 0.1% each of bovine serum albumin, Ficoll, poly(vinylpyrrolidone) and sodium dodecyl sulfate, and 10% dextran sulfate. After each 2 hr hybridization at 53°C, the gel was washed three times for 15 min at 0°C in 6× SSC (0.9 M NaCl, 0.9 M sodium citrate, pH 7.2), frozen overnight at −70°C, washed once for 1 min at 53°C in 6× SSC, and autoradiographed. Autoradiography times were 10, 12, and 4 days for top, middle, and lower panels, respectively. To remove one probe in preparation for hybridization with the next probe, the gel was washed once for 1 min at 65°C in 6× SSC.

pled, the utilization of two pairs of oligonucleotides (one pair for $\beta^{39(UAG)}$-thalassemia, the other pair for IVS 1-110 β^+-thalassemia), re-sulted in an accurate prenatal diagnosis in about 70% of the cases (Thein *et al.*, 1985). This study points out, however, that for a completely comprehensive prenatal diagnosis service, other means of diagnosis (fetal blood sampling) will have to be retained.

An alternative method for detecting single base-pair mutations has recently been described (Myers *et al.*, 1985). The principle of this method lies in the fact that perfectly complementary restriction fragments migrate differently than a heteroduplex on a denaturing gradient polyacrylamide gel system. The method involved hybridizing a genomic DNA sample with a long (428 bases) single-stranded, radioactively labeled probe complementary to the wild-type sequence. After the annealing reaction, the sample was treated with S1 nuclease (to remove unhybridized probe), followed by a restriction enzyme digestion, which reduced the potential heteroduplexes to a uniform size. The DNA sample was then subjected to electrophoresis on a formamide gradient denaturing polyacrylamide gel. In principle, a single base-pair change may lead to difference in the stability of the base-paired, labeled oligomer. The stability difference then results in a migration difference when compared to the wild-type homoduplex on the denaturing gel system. The migration differences were then detected by autoradiography of the gel.

This technique proved successful in the detection of four β-thalassemia point mutations (three transitions, IVS 1-1, IVS 1-6, and IVS 1-110, and one transversion, IVS 1-5). This technique lacks the generality of oligonucleotide hybridization because not all point mutations are discriminated by this technique. The authors estimate that between 25 and 40% of all point mutations can be identified by this technique. They also pointed out that a transition mutation in the first position of the second intervening sequence was not detected with this method. This method should be useful as an initial screen for unknown or uncharacterized β-globin lesions. It has the added value in that small deletions and insertions may be detected since an S1 nuclease step is used. This technique requires about the same amount of DNA as an oligonucleotide hybridization assay; however, since each test destroys the genomic DNA sample, a greater amount of starting material may be required.

VII. CONCLUSION

Since our knowledge or primary protein and DNA sequence information is rapidly expanding, one would expect to find a similar expansion in the use of synthetic oligonucleotides as hybridization probes. In addition to aiding our understanding of specific problems in molecular biology, synthetic oligonucleotides have proven valuable in clinical applications for the diagnosis of human genetic diseases. Since the time and cost of synthesizing DNA oligomers have decreased dramatically in the past decade, the availability of synthetic oligonucleotides should not limit their use. In addition to the techniques mentioned in this review, it is certain that new methodologies will be developed that further expand the present uses of synthetic oligonucleotides.

ACKNOWLEDGMENTS

This work was supported by NIH Grants GM31261 and H129516 to R. B. W. R. B. W. is a member of the City of Hope Cancer Center (NIH CA33572).

REFERENCES

Agarwal, K. L., Brunstedt, J., and Noyes, B. E. (1981). *J. Biol. Chem.* **256,** 1023–1028.
Alwine, J. C., Kemp, D. J., and Stark, G. R. (1977). *Proc. Natl. Acad. Sci. U.S.A.* **74,** 5350–5354.
Anderson, S., and Kingston, I. B. (1983). *Proc. Natl. Acad. Sci. U.S.A.* **80,** 6838–6842.
Antonarakis, S. E., Boehm, C. D., Giardina, P. V. J., and Kazazian, H. H., Jr. (1982). *Proc. Natl. Acad. Sci. U.S.A.* **79,** 137–141.
Astell, C., and Smith, M. (1972). *Biochemistry* **11,** 4114–4120.
Baird, M., Driscoll, C., Schreiner, H., Sciarratta, G., Sansone, G., Niazi, G., Ramirez, F., and Bank, A. (1981). *Proc. Natl. Acad. Sci. U.S.A.* **78,** 4218–4221.
Benton, W. D., and Davis, R. W. (1977). *Science* **196,** 180–182.
Besmer, P., Miller, R. C., Jr., Caruthers, M. H., Kumar, A., Minamoto, K., Van De Sande, J. H. Sidarova, N., and Khorana, H. G. (1972). *J. Mol. Biol.* **72,** 503–522.
Botstein, D., White, R. L., Skolnick, M., and Davis, R. W. (1980). *Am. J. Hum. Genet.* **32,** 314–331.
Chan, S. J., Noyes, B. E., Agarwal, K. L., and Steiner, D. F. (1979). *Proc. Natl. Acad. Sci. U.S.A.* **76,** 5036–5040.
Chang, J. C., and Kan, Y. W. (1981). *Lancet* **2,** 1127–1129.
Chang, J. C., and Kan, Y. W. (1982). *N. Engl. J. Med.* **307,** 30–32.
Chang, S. H., Majumdar, A., Dunn, R., Makabe, O., RajBhandary, U. L., Khorana, H. G., Ohtsuka, E., Tanaka, T., Taniyama, Y. O., and Ikehara, M. (1981). *Proc. Natl. Acad. Sci. U.S.A.* **78,** 3398–3402.
Cheng, C. C., Brownlee, G. G., Carey, N. H., Doel, M. T., Gillam, S., and Smith, M. (1976). *J. Mol. Biol.* **107,** 527–547.
Choo, K. H., Gould, K. G., Rees, D. J. G., and Brownlee, G. G. (1982). *Nature (London)* **299,** 178–180.
Conner, B. J., Reyes, A. A., Morin, C., Itakura, K., Teplitz, R. L., and Wallace, R. B. (1983). *Proc. Natl. Acad. Sci. U.S.A.* **80,** 278–282.
Crea, R., Kraszewski, A., Hirose, T., and Itakura, K. (1978). *Proc. Natl. Acad. Sci. U.S.A.* **75,** 5765–5769.
Davies, K. E., Pearson, P. L., Harper, P. S., Murray, J. M., O'Brien, T., Sarfarazi, M., and Williamson, R. (1983). *Nucleic Acids Res.* **11,** 2303–2312.
De Lange, T., Liu, A. Y. C., Van der Ploeg, L. H. T., Borst, P., Tromp, M. C., and Van Boom, J. H. (1983). *Cell (Cambridge, Mass.)* **34,** 891–900.
Derynck, R., Roberts, A. B., Winkler, M. E., Chen, E. Y., and Goeddel, D. V. (1984). *Cell (Cambridge, Mass.)* **38,** 287–297.
Doel, M. T., and Smith, M. (1973). *FEBS Lett.* **34,** 99–102.
Flavell, R. A., Bernards, R., Kooter, J. M., De Boer, E., Little, P. F. R., Annison, G., and Williamson, R. (1979). *Nucleic Acids Res.* **6,** 2749–2760.
Geever, R. F., Wilson, L. B., Nalleseth, F. S., Milner, P. F., Bittner, M., and Wilson, J. T. (1981). *Proc. Natl. Acad. Sci. U.S.A.* **78,** 5081–5085.

Gillam, S., Waterman, K., and Smith, M. (1975). *Nucleic Acids Res.* **2**, 625–634.

Gitschier, J., Wood, W. I., Goralka, T. M., Wion, K. L., Chen, E. Y., Eaton, D. H., Vehar, G. A., Capon, D. J., and Lawn, R. M. (1984). *Nature (London)* **312**, 326–330.

Goeddel, D. V., Yelverton, E., Ullrich, A., Heyneker, H. L., Miozzari, G., Holmes, W., Seeburg, P. H., Dull, T., May, L., Stebbing, N., Crea, R., Maeda, S., McCandliss, R., Sloma, A., Tabor, J. M., Gross, M., Familletti, P. C., and Pestka, S. (1980a). *Nature (London)* **287**, 411–416.

Goeddel, D. V., Shepard, H. M., Yelverton, E., Leung, D., and Crea, R. (1980b). *Nucleic Acids Res.* **8**, 4057–4074.

Goossens, M., Dumez, Y., Kaplan, L., Lupker, M., Chabret, C., Henrion, R., and Rosa, J. (1983). *N. Engl. J. Med.* **309**, 831–833.

Grantham, R., Gautier, C., and Gouy, M. (1980). *Nucleic Acids Res.* **8**, 1893–1912.

Gray, P. W., Aggarwal, B. B., Benton, C. V., Bringman, T. S., Henzel, W. J., Jarrett, J. A., Leung, D. W., Moffat, B., Ng, P., Svedersky, L. P., Palladino, M. A., and Nedwin, G. E. (1984). *Nature (London)* **312**, 721–724.

Grunstein, M., and Hogness, D. S. (1975). *Proc. Natl. Acad. Sci. U.S.A.* **72**, 3961–3965.

Gubler, U., Kilpatrick, D. L., Seeburg, P. H., Gage, L. P., and Udenfriend, S. (1981). *Proc. Natl. Acad. Sci. U.S.A.* **78**, 5484–5487.

Gusella, J. F., Wexler, N. S., Conneally, P. M., Naylor, S. L., Anderson, A. E., Tanzi, R. E., Watkins, P. C., Ottina, K., Wallace, M. R., Sakaguchi, A. Y., Young, A. B., Shoulston, I., Bonilla, E., and Martin, J. B. (1983). *Nature (London)* **306**, 234–238.

Guyaux, M., Cornelissen, A. W. C. A., Pays, E., Steinert, M., and Borst, P. (1985). *EMBO J.* **4**, 995–998.

Houghton, M., Stewart, A. G., Doel, S. M., Emtage, J. S., Eaton, E. A., Smith, J. C., Patel, T. P., Lewis, H. M., Porter, A. G., Birch, J. R., Cartwright, T., and Carey, N. H. (1980). *Nucleic Acids Res.* **8**, 1913–1931.

Hudson, P., Haley, J., Cronk, M., Shine, J., and Niall, H. (1981). *Nature (London)* **291**, 127–131.

Hutchinson, C. A., III, Phillips, S., Edgell, M. H., Gillam, S., Jahnke, P., and Smith, M. (1978). *J. Biol. Chem.* **253**, 6551–6560.

Itakura, K., Hirose, T., Crea, R., and Riggs, A. D. (1977). *Science* **198**, 1056–1063.

Itakura, K., Rossi, J. J., and Wallace, R. B. (1984). *Annu. Rev. Biochem.* **53**, 323–356.

Jaye, M., De La Salle, H., Schamber, F., Balland, A., Kohli, V., Findeli, A., Tolsteshev, P., and Lecocq, J.-P. (1983). *Nucleic Acids Res.* **11**, 2325–2335.

Kan, Y. W., and Dozy, A. M. (1978). *Lancet* **2**, 910–912.

Kan, Y. W., and Dozy, A. M. (1980a). *Proc. Natl. Acad. Sci. U.S.A.* **75**, 5631–5635.

Kan, Y. W., and Dozy, A. M. (1980b). *Science* **209**, 388–391.

Kan, Y. W., Golbus, M. S., and Dozy, A. M. (1976). *N. Engl. J. Med.* **295**, 1165–1167.

Kan, Y. W., Dozy, A. M., Stamatoyannopoulos, G., Hadjiminas, M. G., Zachariades, Z., Furbetta, M., and Cao, A. (1979). *Blood* **54**, 1434–1438.

Kan, Y. W., Lee, K. Y., Furbetta, M., Angius, A., and Cao, A. (1980). *N. Engl. J. Med.* **302**, 185–188.

Kazazian, H. H., Jr. (1985). *Trends NeuroSci. (Pers. Ed.)* **8**, 192–200.

Kidd, V. J., Wallace, R. B., Itakura, K., and Woo, S. L. C. (1983). *Nature (London)* **304**, 230–304.

Kidd, V. J., Golbus, M. S., Wallace, R. B., Itakura, K., and Woo, S. L. C. (1984). *N. Engl. J. Med.* **310**, 639–642.

Lai, C.-J., Markoff, L. J., Zimmerman, S., Cohen, B., Berndt, J. A., and Chanock, R. M. (1980). *Proc. Natl. Acad. Sci. U.S.A.* **77**, 210–214.

Lathe, R. (1985). *J. Mol. Biol.* **183**, 1–12.

Legon, S., Glover, D. M., Hughes, J., Lowry, P. J., Rigby, P. W. J., and Watson, C. J. (1982). *Nucleic Acids Res.* **10,** 7905–7918.

Mears, J. G., Ramirez, F., Leibowitz, D., and Bank, A. (1978). *Cell (Cambridge, Mass.)* **15,** 15–23.

Mevarech, M., Noyes, B. E., and Agarwal, K. L. (1979). *J. Biol. Chem.* **254,** 7472–7475.

Miyada, C. G., Klofelt, C., Reyes, A. A., McLaughlin-Taylor, E., and Wallace, R. B. (1985). *Proc. Natl. Acad. Sci. U.S.A.* **82,** 2890–2894.

Montgomery, D. L., Hall, B. D., Gillam, S., and Smith, M. (1978). *Cell (Cambridge, Mass.)* **14,** 673–680.

Myers, R. M., Lumelsky, N., Lerman, L. S., and Maniatis, T. (1985). *Nature (London)* **313,** 495–498.

Nickoloff, J. A., and Hallick, R. B. (1982). *Nucleic Acids Res.* **10,** 8191–8210.

Niyogi, S. K., and Thomas, C. A., Jr. (1968). *J. Biol. Chem.* **243,** 1220–1223.

Noda, M., Takahashi, H., Tanabe, T., Toyosato, M., Furutani, Y., Hirose, T., Asai, M., Inayama, S., Miyata, T., and Numa, S. (1982). *Nature (London)* **299,** 793–797.

Noyes, B. E., Mevarech, M., Stein, R., and Agarwal, K. L. (1979). *Proc. Natl. Acad. Sci. U.S.A.* **76,** 1770–1774.

Oehme, R., Kohne, E., Kleihauer, E., and Horst, J. (1983). *Hum. Genet.* **64,** 376–379.

Ohkubo, H., Kageyama, R., Ujihara, M., Hirose, T., Inayama, S., and Nakanishi, S. (1983). *Proc. Natl. Acad. Sci. U.S.A.* **80,** 2196–2200.

Omiecinski, C. J., Walz, F. G., Jr., and Vlasuk, G. P. (1985). *J. Biol. Chem.* **260,** 3247–3250.

Orkin, S. H., Alter, B. P., Altay, C., Mahoney, M. J., Lazarus, H., Hobbins, J. C., and Nathan, D. G. (1978). *N. Engl. J. Med.* **299,** 166–172.

Orkin, S. H., Old, J., Lazarus, H., Altay, C., Gurgey, A., Weatherall, D. J., and Nathan, D. J. (1979a). *Cell (Cambridge, Mass.)* **17,** 33–42.

Orkin, S. H., Old, J. M., Weatherall, D. J., and Nathan, D. G. (1979b). *Proc. Natl. Acad. Sci. U.S.A.* **76,** 2400–2404.

Orkin, S. H., Little, P. F. R., Kazazian, H. H., Jr., and Boehm, C. D. (1982). *N. Engl. J. Med.* **307,** 32–36.

Orkin, S. H., Markham, A. F., and Kazazian, H. H., Jr. (1983). *J. Clin. Invest.* **71,** 775–779.

Ottolenghi, S., Comi, P., Giglioni, B., Tolstoshev, P., Lanyon, W. G., Mitchell, G. J., Russo, G., Musumeci, S., Schiliro, G., Tsistrakis, G. A., Charache, S., Wood, W. G., Clegg, J. B., and Weatherall, D. J. (1976). *Cell (Cambridge, Mass.)* **9,** 71–80.

Panny, S. R., Scott, A. F., Smith, K. D., Phillips, J. A., III, Kazazian, H. H., Jr., Talbot, C. C., Jr., and Boehm, C. D. (1981). *Am. J. Hum. Genet.* **33,** 25–35.

Pennica, D., Nedwin, G. E., Hayflick, J. S., Seeburg, P. H., Derynck, R., Palladino, M. A., Kohr, W. J., Aggarwal, B. B., and Goeddel, D. V. (1984). *Nature (London)* **312,** 724–729.

Phillips, J. A., III, Scott, A. F., Smith, K. D., Young, K. E., Lightbody, K. L., Jiji, R. M., and Kazazian, H. H., Jr. (1979a). *Blood* **54,** 1439–1445.

Phillips, J. A., III, Scott, A. F., Kazazian, H. H., Jr., Stetten, G., and Thomas, G. H. (1979b). *Johns Hopkins Med. J.* **145,** 57–60.

Phillips, J. A., III, Panny, S. R., Kazazian, H. H., Jr., Boehm, C. D., Scott, A. F., and Smith, K. D. (1980). *Proc. Natl. Acad. Sci. U.S.A.* **77,** 2853–2856.

Pirastu, M., Kan, Y. W., Cao, A., Conner, B. J., Teplitz, R. L., and Wallace, R. B. (1983). *N. Engl. J. Med.* **309,** 284–287.

Pirastu, M., Del Senno, L., Conconi, F., Vullo, C., and Kan, Y. W. (1984a). *Nature (London)* **307,** 76.

Pirastu, M., Kan, Y. W., Galanello, R., and Cao, A. (1984b). *Science* **223,** 929–930.

Rabbitts, T. H., Forster, A., Smith, M., and Gillam, S. (1977). *Eur. J. Immunol.* **7,** 43–48.

Razin, A., Hirose, T., Itakura, K., and Riggs, A. D. (1978). *Proc. Natl. Acad. Sci. U.S.A.* **75,** 4268–4270.

Reyes, A. A., and Wallace, R. B. (1984). *Genet. Eng.* **6,** 157–173.

Reyes, A. A., Johnson, M. J., Schold, M., Ito, H., Ike, Y., Morin, C., Itakura, K., and Wallace, R. B. (1981). *Immunogenetics* **14,** 383–392.

Rosatelli, C., Falchi, A. M., Tuveri, T., Scalas, M. T., Di Tucci, A., Monni, G., and Cao, A. (1985). *Lancet* **2,** 241–243.

Sasavage, N. L., Smith, M., Gillam, S., Astrell, C., Nilson, J. H., and Rottman, F. M. (1980). *Biochemistry* **19,** 1737–1743.

Scheller, R. H., Dickerson, R. E., Boyer, H. W., Riggs, A. D., and Itakura, K. (1977). *Science* **196,** 177–180.

Smith, M. (1983). *In* "Methods of RNA and DNA Sequencing" (S. M. Weissman, ed.), pp. 23–68. Praeger, New York.

Sood, A. K., Pereira, D., and Weissman, S. M. (1981). *Proc. Natl. Acad. Sci. U.S.A.* **78,** 616–620.

Southern, E. M. (1975). *J. Mol. Biol.* **98,** 503–517.

Steinmetz, M., Winoto, A., Minard, K., and Hood, L. (1982). *Cell (Cambridge, Mass.)* **28,** 489–498.

Stetler, D., Das, H., Nunberg, J. H., Saiki, R., Sheng-Dong, R., Mullis, K. B., Weissman, S. M., and Erlich, H. A. (1982). *Proc. Natl. Acad. Sci. U.S.A.* **79,** 5966–5970.

Stewart, J. W., and Sherman, F. (1974). *In* "Molecular and Environmental Aspects of Mutagenesis" (L. Prakesh, F. Sherman, M. W. Miller, C. W. Lawrence, and H. W. Taber, eds.), pp. 102–107. Thomas, Springfield, Illinois.

Studencki, A. B., and Wallace, R. B. (1984). *DNA* **3,** 7–15.

Studencki, A. B., Conner, B. J., Impraim, C. C., Teplitz, R. L., and Wallace, R. B. (1985). *Am. J. Hum. Genet.* **37,** 42–51.

Suggs, S. V., Wallace, R. B., Hirose, T., Kawashima, E. H., and Itakura, K. (1981). *Proc. Natl. Acad. Sci. U.S.A.* **78,** 6613–6617.

Szostak, J. W., ʹiles, J. I., Bahl, C. P., and Wu, R. (1977). *Nature (London)* **265,** 61–63.

Szostak, J. W., Stiles, J. I., Tye, J.-K., Chiu, P., Sherman, F., and Wu, R. (1979). *In* "Methods in Enzymology" (R. Wu, ed.), Vol. **68,** pp. 419–428. Academic Press, New York.

Thein, S. L., Wainscoat, J. S., Old, J. M., Sampietro, M., Fiorelli, G., Wallace, R. B., and Weatherall, D. J. (1985). *Lancet* **2,** 345–347.

Ullrich, A., Gray, A., Berman, C., and Dull, T. J. (1983). *Nature (London)* **303,** 821–825.

Ullrich, A., Gray, A., Wood, W. I., Hayflick, J., and Seeburg, P. H. (1984a). *DNA* **3,** 387–392.

Ullrich, A., Coussens, L., Hayflick, J. S., Dull, T. J., Gray, A., Tam, A. W., Lee, J., Yarden, Y., Libermann, T. A., Schlessinger, J., Downard, J., Mayes, E. L. V., Whittle, N., Waterfield, M. D., and Seeburg, P. H. (1984b). *Nature (London)* **309,** 418–425.

Ullrich, A., Berman, C. H., Dull, T. J., Gray, A., and Lee, J. M. (1984c). *EMBO J.* **3,** 361–364.

Ullrich, A., Bell, J. R., Chen, E. Y., Herrera, R., Petruzzelli, L. M., Dull, T. J., Gray, A., Coussens, L., Liao, Y.-C., Tsubokawa, M., Mason, A., Seeburg, P. H., Grunfeld, C., Rosen, O. M., and Ramachandran, J. (1985). *Nature (London)* **313,** 756–761.

Villa-Komaroff, L., Efstratiadis, A., Broome, S., Lomedico, P., Tizard, R., Naber, S. P., Chick, W. L., and Gilbert, W. (1978). *Proc. Natl. Acad. Sci. U.S.A.* **75,** 3727–3731.

Wallace, R. B., and Itakura, K. (1984). *In* "Solid Phase Biochemistry: Analytical Synthetic Aspects" (W. H. Scouten, ed.), pp. 631–663. Wiley, New York.

Wallace, R. B., Shaffer, J., Murphy, R. F., Bonner, J., Hirose, T., and Itakura, K. (1979). *Nucleic Acids Res.* **6,** 3543–3557.

Wallace, R. B., Johnson, P. F., Tanaka, S., Schold, M., Itakura, K., and Abelson, J. (1980). *Science* **209,** 1396–1400.

Wallace, R. B., Johnson, M. J., Hirose, T., Miyake, T., Kawashima, E. H., and Itakura, K. (1981). *Nucleic Acids Res.* **9,** 879–894.

Watson, J. D., and Crick, F. C. (1953). *Nature (London)* **171,** 737–738.

Whitehead, A. S., Goldberger, G., Woods, D. E., Markham, A. F., and Colten, H. R. (1983). *Proc. Natl. Acad. Sci. U.S.A.* **80,** 5387–5391.

Wilson, J. M., Frossard, P., Nussbaum, R. L., Caskey, C. T., and Kelley, W. N. (1983a). *J. Clin. Invest.* **72,** 767–772.

Wilson, J. M., Young, A. B., and Kelley, W. N. (1983b). *N. Engl. J. Med.* **309,** 900–910.

Winter, G., Fields, S., and Gait, M. J. (1981). *Nucleic Acids Res.* **9,** 237–245.

Wood, W. I., Capon, D. J., Simonsen, C. C., Eaton, D. L., Gitschier, J., Keyt, B., Seeburg, P. H., Smith, D. H., Hollingshead, P., Wion, K. L., Delwart, E., Tuddenham, G. D., Vehar, G. A., and Lawn, R. M. (1984). *Nature (London)* **312,** 330–337.

Wood, W. I., Gitschier, J., Lasky, L. A., and Lawn, R. M. (1985). *Proc. Natl. Acad. Sci. U.S.A.* **82,** 1585–1588.

Woods, D. E., Markham, A. F., Ricker, A. T., Goldberger, G., and Colten, H. R. (1982). *Proc. Natl. Acad. Sci. U.S.A.* **79,** 5661–5665.

Wu, R. (1972). *Nature (London) New Biol.* **236,** 198–200.

Wu, R., Tu, C.-P. D., and Padmanabhan, R. (1973). *Biochem. Biophys. Res. Commun.* **55,** 1092–1099.

Index

A

Acetic anhydride, with N,N-dimethyl aminopyridine, as capping reagent, 56, 57

N-Acyl nucleoside, preparation, 116, 117

Adenosine
  amino group, protection, 116
  in higher-order structure, 174
  modification reaction, 166–168

Agarose gel electrophoresis, 190, 203

2-Amino-2'-deoxyadenosine, 34, 35

Ammonia deprotection, side reaction, 33–36

Ammonium hydroxide, 55, 65, 74, 75

Amniocentesis, 217

Angiotensin I, gene synthesis, 108

Angiotensinogen, 211

Antibody screening, 209

Antihemophilic factor IX, 212, 213

α_1-Antitrypsin, 219

Antiviral drug, 85

Arenesulfonyl azolide, 124

Arenesulfonyl chlorides, 13

Arene-sulfonyl chloride–methylimidazole method, 29

Arenesulfonyl-3-nitro-1,2,4-triazoles, 13

Arenesulfonyl tetrazoles, 13

Aryl sulfonyl chloride, 4, 5

[α-^{32}P]ATP, 165

[γ-^{32}P]ATP, 166

Automation, 32, 73, 74
  chemistry, 40, 41
  criteria, 39, 40
  hardware, 41, 42
  software, 41
  valves, 42

B

Bacterial alkaline phosphatase, 166, 189, 196

Bacteriorhodopsin gene, 209

BAL 31 nuclease, 149, 151

Base, nonstandard, 34–36

Benzenesulfonic acid, 24, 25, 28

2'(3')-O-Benzoyluridine, as universal linker, 54, 55

Benzyl ether, 119

2-Benzylsulfonylethyl linkage, 18, 20

BioLogicals, synthesizer, 40

Biotin labeling, 87

Bisaminophosphine, 67, 68

Bond, *see also* Phosphoramidite internucleotidic linkage, internucleotidic, formation, 29, 30

Bradykinin, gene synthesis, 108

BSE, *see* 2-Benzylsulfonylethyl linkage

tert-Butyldimethylsilyl group, 120, 126

C

CAMET, *see* 2-(4-Carboxyphenylmercapto)ethanol

Capping, 28, 30–32
  need for, 56, 57, 71